Peering Through
The Bushes

Peering Through The Bushes

A Commentary by Nationally Syndicated Environmental Columnist Edward Flattau

Edward Flattau

Copyright © 2004 by Edward Flattau.

Library of Congress Number: 2004094362
ISBN : Hardcover 1-4134-6113-1
Softcover 1-4134-6112-3

All rights reserved. No part of this book may be reproduced or transmitted in any form or by any means, electronic or mechanical, including photocopying, recording, or by any information storage and retrieval system, without permission in writing from the copyright owner.

This book was printed in the United States of America.

To order additional copies of this book, contact:
Xlibris Corporation
1-888-795-4274
www.Xlibris.com
Orders@Xlibris.com

25236

This Book Is Dedicated To
The Amercan Voter In 2004
And beyond

PREVIOUS WORKS BY THE AUTHOR

Tracking the Charlatans
(Global Horizons Press, 1998)

Evolution of a Columnist
(Global Horizons Press, 2003)

Contents

Introduction 13

Chapter One: Dear Diary 17
    2001: The First Year of a Presidency
    2002: The Beat Goes On
    2003: . . . And On
    2004: Coda

Chapter Two: Setting The Stage 42
    Dual Personalities
    Closet Environmentalist?
    Short-Lived Honeymoon
    The Isolationist Tinge
    Painless War
    I've Got A Secret
    The Legacy

Chapter Three: Texas True And Blue 48
    Only So Far
    The Platform Tell All
    In His Words
    Uniter Or Divider?
    Prophetic

Chapter Four: The Early White House Days 55
Anything But Clinton
How Dumb Can You Get?
Translucent Compassion
Early Pluses And Minuses
James Watt In Skirts
Thoughts Unspoken
W's Brave New World
Anyone But Clinton
Earth Day Déjà Vu

Chapter Five: Arrogance And Unilateralism 63
Forget Nuance
The Ugly American
The Not So Grand Obsession
Alieanation
To Be, Or Not To Be
Image Making
Poison Pen Letter
Above The Law
Moral Clarity
Tomfoolery
Let The U.N. In
Humility Or Humiliation?

Chapter Six: Stealth And Snow Jobs 75
Loading Up For Bear
Justice Triumphs
Overthrust Malarkey
Specious Semantics
Arctic Candor
Anatomy Of A Scam
Trapped
Bad Storytelling
Operation Stealth
Sins Of Omission

In The Eyes Of The Beholder
Dollar Deception
The Patriotism Factor
Trust Us
Summer Doldrums
Transparency

Chapter Seven: Blunders And Blind Spots—A Prisoner Of Ideology 92

The American Way
A Shaky Pillar
Justice Be Served
Beyond Ideology
Timidity From A Tough Guy
It's The Science, Stupid!
Cockeyed Priorities
The False Populist
Volunteerism Copout?
Health Versus Profit
Denial
Your Kinda Guy?
Gagging On A Rule
The Web Of Life
The Bush Rationale

Chapter Eight: Worse Than Reagan? 107

Divide And Conquer
Hoodwinking The Clergy
Campaign Potholes
Under Suspicion
Poll Watching
Terrorizing The Environment
Supreme Irony
Game Plan
Abuse Of The Year
Is Anybody Listening?

    Secret To Success
    Fall From Grace
    The Worst Ever?

Chapter Nine: Where To From Here? *125*
    Future Challenges
    An Ecumenical Test
    Wild Card
    The World Under Bush
    A Question Of Values
    Power Failure
    The Vision Thing
    Not Getting It
    Unraveling The Base?
    Consumed With Consumption
    Signs Of Hope
    Earth Day 2004
    Credit Where Credit Is Due

Notes *139*

Introduction

Peering Through the Bushes is a running commentary on George W. Bush's relationship to environmental issues. Rather than rendering a detailed documentation of the president's performance, the book conveys a broad impressionistic picture of the causes, impacts, and implications of Bush's controversial environmental policies. The author is the nation's senior nationally syndicated environmental columnist and has been following Bush's environmental policy-making closely enough over the years to feel comfortable making subjective judgments about the president's character and motivation on those matters.

Not surprisingly, many of Bush's environmental positions mirror those of his father. But the son strays quite a bit further from the mainstream than Bush Senior ever did. Junior is much more enamored of conservative ideology than his father was, and is seemingly even less convinced of the imminence and gravity of environmental threats.

You can thus understand why George W. Bush's takeover of the presidency has caused environmentalists to greet the 21st Century with trepidation. They watched with dismay as their issues were undermined by a discredited conservative ideological approach they thought had been put to rest when Ronald Reagan left office 15 years earlier.

Suddenly, states' rights, corporate volunteerism, and marketplace incentives were in vogue. Stringent environmental regulation and primacy of federal laws over local ones were out of favor except on the rarest of occasions. Bush gravitated towards an anachronistic unilateral stance on any number of international environmental matters, despite the United States inevitably having to operate in an ever more environmentally interdependent world. His "new" environmental approach catapulted the nation back in time to the Reagan years, with their embarrassing and sometimes alarming ventures into isolationism.

This was not an auspicious way to inaugurate a century in which humanity can be expected to either restore an ecologically stressed planet to good health or doom earth to an irreversible downward environmental spiral.

The specious rhetoric heard so often during Reagan's reign was once again emanating from the White House. Preferential treatment in environmental controversies was granted to business interests in the guise of returning some sense of "balance" to natural resources protection. Rationalizations were concocted for opening up previously protected public conservation areas to industrial activity. Brimming with animus at his predecessor's ideological stance, sexual dalliances, and defeat of his father, George W. Bush was obsessed with rescinding as many of Clinton's environmental initiatives as possible.

Even after his inauguration, Bush displayed no sign of being able to broaden his environmental perspective beyond preconceived, narrowly drawn ideological nostrums. Those solutions might be applicable on occasion, but no more than that. Employing the politically correct methodology often appeared more important to Bush than reaching the correct result in environmental controversies. In the wake of the World Trade Center attack, he exhibited no recognition that reversing environmental degradation was a major component of any long-term solution to terrorism.

In short, Bush's handling of environmental issues in public office has been a tale of deception, delusion, intractable doctrinaire thinking, indifference, and regulatory rollbacks.

Chapter One contains entries in a diary that this author began when the president took office. As you will see, Bush engaged in his ideological vendetta against existing environmental safeguards from the get-go.

Chapter One

Dear Diary

2001: The First Year of a Presidency

March 21: Bush's disdain for Clinton, combined with a limited vision of environmental protection and an eagerness to cater to his energy industry cronies, has led him to seek rollbacks of environmental regulations.

March 22: Bush and his people are so incensed at Clinton's last-minute environmental directives that they seem determined to reverse them, regardless of the proposals' popularity (or merit). *They are tempting fate with such lapses in both political and ethical judgment.*

April 17: Bush doesn't seem to appreciate the deep reservoir of public support for environmental protection. He has been sloppy, cavalier, and uninformed about environmental issues early in his tenure, largely because he believes they can easily be papered over. His myopia stems from having been able either to mollify environmental critics or sweep their concerns under the rug when governor of Texas. *He soon will discover that Washington is not as malleable.*

April 18: "Sound science", "balance", and "sensible regulation" are code words for right-wing attempts to dismantle our environmental regulatory infrastructure.

... Bush tries to blunt a bad first impression on the eve of Earth Day by backing away from canceling a Clinton Administration arsenic regulation, and issuing directives upholding wetland protection and disclosure of lead contamination.

April 19: I believe Bush and the conservative ideologues who are his environmental advisors are misguided rather than malevolent.

April 20: Will Bush's exposure to more intense, widespread public environmental opposition than he experienced as governor of Texas persuade him to shift away from doctrinaire rigidity and towards non-ideological, pragmatic environmental policy? Or will he stick to his parochial guns and end up in seclusion, isolated from all but his hard-core ultra-right-wing supporters?

April 21: Will Bush realize at some point that the majority of the country opposes his philosophy of devolution and market-based incentives as the foundations of environmental policy? If so, will he alter his policies at the risk of alienating his hard-core ultra Right Wing supporters?
  ... On the eve of Earth Day, Bush in his weekly radio address made no reference to the environment.
  ... Bush clearly has no intellectual affinity for environmental concerns and appears to be relying on ideological advisors who divide their time between reversing popular Clinton policies and warding off unfavorable publicity through the spread of misinformation that fosters confusion.
  ... The Dutch environmental minister commenting on Bush's pronouncement that the Kyoto global warming treaty is dead: "No one country can unilaterally declare an international treaty null and void." *That's what you think!*

April 22: Bush at the Summit of Americas in Quebec City: "I understand the need for environmental codicils, but we should not allow them to destroy the spirit of free trade." *Did it ever occur to him that free trade agreements that defile the environment might not be worth spit?*

April 24: Bush: "We're going to make decisions based upon sound science, not some environmental fad or what may sound good." *What is he talking about?*

April 25: Bush believes everything his environmental advisors tell him because he is a true believer. To reject their counsel would not only cause him to repudiate trusted aides (on a subject in which he has little interest or expertise) but force him to challenge the core set of ideological convictions that anchor his political life.

. . . Bush: "I think that the public will come to realize that I am a person who believes in a common-sense approach to the environment, that I strongly believe that technologies will help us achieve a common objective, which is cleaner air and cleaner water." *The environmental community also believes in these things. It just finds Bush's common-sense approach "nonsensical."*

April 26: Bush calls for closer cooperation between federal authorities and the local populace, which is fine and dandy. But he is talking about local management and control of federal lands that belong to all Americans. Local populations by virtue of their proximity may deserve to be consulted first and enjoy easier access, but they should have no greater say on the ultimate disposition and use of the land than the rest of their countrymen.

May 1: Bush is so deeply indoctrinated in conservative ideology Texas-style that he seems unswerving in his assumption that what he has been taught is right. One wonders if he is capable of ever conceding he is wrong. Can he exchange ideology for pragmatism and adopt a mix of policies that work, or will his presidency be an exercise in toeing a rigid line out of ideological correctness?

May 7: Bush's use of the epithet "environmental extremists" to describe his opponents amounts to a blanket indictment of a majority of Americans.

May 14: If Bush's rationale for lowering income taxes is to provide SUV owners with additional money to ease the burden of the higher price of the gasoline that their vehicles guzzle, what does that say about his vision for the nation's energy future?

May 16: Bush's first four months in office have been a disaster. He has alienated our longstanding allies, stirred up our enemies, rolled back environmental laws, set the stage for the industrial rape of public lands, and turned back the clock on development of an environmentally sustainable energy policy.

May 18: Bush appears to lack the intellectual curiosity to challenge and conceivably transcend the political philosophy on which he was weaned.

May 19: In his radio address today, Bush said, "it's time to leave behind rancorous old arguments and build a positive new consensus (on energy policy)." *Mr. President, disagreement is not rancorous!*

June 7: Bush and his minions make certain that a disclaimer always accompanies their environmentally destructive initiatives. *We need to pay more attention to these disclaimers in order to expose their diversionary intent.*

June 10: Bush is just a front man for his ideologically driven advisors. He believes fervently in what he is told, and gives no indication of engaging in analytical questioning or possessing the intellectual energy to change course.

 . . . A Bush Administration official says Europeans think of the president as a "shallow, arrogant, gun-loving, abortion-hating, Christian fundamentalist, Texan buffoon, and hard-line unilateralist." *The "Texas buffoon" epithet is over the top, but otherwise, the Europeans may be on to something.*

 . . . From Frank Rich in 6/9/01 *New York Times*, regarding clues to Bush's behavior: "CEOs are used to flying their own planes, seeing only their own subordinates, and being accountable to no one. They are profoundly certain of their own value system. They have contempt for the public and the press. They have none of the accountability required of a president of the United States." *Rich may be on to something.*

June 13: I wouldn't go as far as to call Bush a simpleton, but his first trip to Europe reinforces the initial impression that he has a very narrow cognitive range. He can only repeat the same stock phrases over and over again, and you get the impression that if he couldn't decipher the print in his prepared script, he would be at a loss for words.

June 14: The media is putting the best face on a guy who in reality is an out-and-out disaster.

June 18: To fend off criticisms he is acting unilaterally, Bush declares, "A unilateralist doesn't sit around tables listening to the views of others." *Yes, he does, if he wants to deceive others into thinking he is taking their concerns seriously.*

... Bush is so totally convinced of the righteousness of his policies that he wouldn't dream of modifying them. When he talks about his openness in consulting the Europeans who disagree with him, it's all about trying to con them through feigned attentiveness.

June 27: Bush is not inherently malevolent. He just believes his conservative orthodoxy is best, and he is doing the right thing by delegating federal authority to local jurisdictions and letting market forces rather than regulation advance the cause of environmental protection.

July 1: Bush may be stymied in enacting unpopular environmental reforms, but his evangelical orientation will never permit him to cease and desist.

July 4: I think Interior Secretary Gale Norton is a consummate con artist. She hasn't renounced her right-wing fringe ideology one iota. All she does in front of congressional interrogators is flash a toothy grin, spout environmentally pleasing rhetoric, and commit to nothing.

July 10: Bush is intellectually either unwilling or unable to venture beyond a few basic conservative principles, no matter what environmental situation he confronts. *My take is it is a combination of the two causative factors.*

July 15: Bush has a fanatic fealty to free-market economics to the exclusion of alternative approaches, including government subsidies to stimulate the economy, unless of course the subsidies are directed to the president's corporate buddies.

July 16: In the six months that Bush has been in office, he has managed to make us one of the most despised nations in the world.

July 24: Bush keeps promising he will soon come up with an alternative to the Kyoto global warming treaty, but everyone already knows what to expect. His proposal will be a mix of market incentives and voluntary measures that won't come close to satisfying the rest of the world.

July 25: Bush's pledge to be a uniter rather than a divider can only be viewed as a bad joke. During his first six months, he has alienated our European allies and created one of the most fractious partisan atmospheres ever witnessed in the nation's capital.

August 25: Bush marvels at nature on his 1500-acre Crawford, Texas, ranch. But in his simple-minded embrace of conservative ideology, he is unable to extend his reverence for nature to the undeveloped public lands that his drill-and-dredge policies would ravage.

August 28: With the exception of the National Park Service director, Bush's top environmental appointees all were recruited from high posts in polluting industries. Can these individuals rise above previous loyalties, and will they even need to?

September 8: In many respects, Bush's treatment of the environment in his first eight months in office is a carbon copy of Ronald Reagan's policy 20 years ago. From this similarity, one can deduce that Bush has politically written off the environmental community in the belief he can win without it. We shall see.

September 14: The massive acts of terrorism against our country give Bush some respite over such knotty issues as the environment. Will these momentous crimes in any manner change the way he looks at things, including his rather laissez-faire approach to protecting the environment?

September 16: Will Bush have enough sense to recognize that environmental protection is a universal language with the potential to transcend religious, cultural, ethnic, political, and even economic differences and create a common bond between the nations of the world?

September 17: The Bush Administration is going to have to swallow its pride, ditch its arrogant unilateral stance, and come hat in hand to nations of the world seeking their cooperation in the war against the shadowy, ubiquitous foe of terrorism. *What irony!*

September 19: Will Bush exploit the terrorism crisis by using it as a pretext for rolling back environmental protections that he could not budge before the Twin Tower attacks? He would make a grave mistake if he did so, because our survival hinges every bit as much on how we handle environmental threats as terrorist ones.

October 2: In the aftermath of the terrorist attack, a further irony is that the "States' Rights" president is going to end up presiding over the greatest expansion of the federal government in decades. Will this carry over in prudent fashion to the regulation of the environment? *I doubt it.*

October 4: Fox media commentator Tony Snow declared that one positive thing resulting from the terrorist attack was that it focused the country on "really important matters" and shunted aside such "idiotic trivia" as global warming. That seems to be the Hard Right's myopic view. Will Bush take the same position?

October 11: There are hints that Bush will remain captive to hard-core conservative ideology on domestic matters, even as he matures on foreign policy under the press of the terrorist threat. The hints include hesitation to expand the government's domestic role, even in the case of federalizing airport security personnel, and dogmatic insistence on further tax cuts, even as military demands on the national budget proliferate exponentially.

October 12: Bush has just announced he is sticking to the party line of drilling in the Arctic National Wildlife Refuge and any other public lands where he can gain access for industrial development. *I guess once an oil man, always an oil man.*

November 4: Religion is an essential part of Bush's life. Yet his environmental policies frequently put him at odds with the theological leaders of the world's major religions.

November 17: Bush's ideological minions in federal environmental agencies are attempting to carry out as unobtrusively as possible their pro-industry, anti-regulatory agenda, undoubtedly with the hope that their activities will go unnoticed by a public preoccupied with terrorism threats.

November 26: Bush displayed a high degree of cockiness prior to entering office, and it has only grown as his approval rating has soared in response to his tough stand against terrorism. This cockiness has buttressed his unshakable conviction in the righteousness of his decision-making, and fueled his disdain of Congress, the media, and most of all, criticism. It's this cockiness that could be his Achilles' heel if events turn sour and his high approval rating begins to slide. Will he ever be able to handle, or even recognize, when he is off course and in danger of overstepping his mandate?

November 29: Bush has innate shrewdness but limited intellectual breadth. Whether the latter is congenital or merely a matter of laziness is unclear.

December 4: I don't think Bush has the intellectual elasticity to voluntarily break out of his conservative box when presented with environmental options that would seem the pragmatic thing to do in the national interest. Yet his limited focus is a virtue in the war on terrorism, which does demand a high degree of single-mindedness—and who knows, the distraction might spare environmental activists some unpleasant moments.

December 15: Bush and his subordinates are brimming with arrogance, undoubtedly boosted by the so far successful prosecution of the war against terrorism. In particular, the cockiness is manifested by frequently high-handed treatment of the international community. In regard to the domestic scene, it would be wise for them not to make any more enemies in Congress than they have to, because if the political climate should deteriorate, the Bush team will need every friend it can get.

2002: *The Beat Goes On*

February 12: Bush is trying to parlay his popularity derived from prosecuting the war on terrorism into justification for advancing a domestic ideological agenda inimical to a majority of Americans (e.g., reducing the size of government and shifting the responsibilities from canceled federal programs to volunteerism and the "good graces" of the marketplace). This will eventually backfire. I'm just not clear when.

February 13: Bush is smart enough to remain silent and let his lackeys make the insinuation that if you disagree with his domestic policies, you are jeopardizing the existence of a united front needed to combat terrorism and are raising questions about your patriotism.

February 20: Flush with the United States victory in Afghanistan, Bush is displaying definite signs of hubris, a sense of invincibility fueled by record high public approval ratings. If Shakespeare is any indicator, our president is riding for a fall.

February 21: The White House characterized Bush's public saber-rattling as refreshing plain talk. I would describe it as arrogant simple-mindedness. Foreign diplomacy is a subtle art that adjusts to different cultures and values as the need arises. If tough talk is necessary, the message should usually be delivered in private. *But then, what would you expect? Bush himself boasts that he "doesn't do nuance."*

. . . Bush's belligerent public tone may play well in the hills of West Texas, but it creates an image more of a bully than a benefactor among many in the capitals of our allies as well as our foes. If this keeps up, Americans abroad are going to have an increasingly difficult time.

March 25: It's ironic that a president who wears moral clarity on his sleeve chooses to violate the ethical imperative of providing the American people the information they need to weigh the pros and cons of what is being asked of them in regard to the determination of their destiny.

April 22: Bush says he has the same goals as the environmental movement. Only the methodology is different. A carrot works better than a stick. The problem is that the stick has a very good track record, while the carrot has a very spotty one. Where environmental laws enacted over the past 30 years have not met expectations, the explanation usually lies with half-hearted or negligible enforcement. By contrast, Bush's voluntary approach—sweetened by monetary incentives—has had uneven results. If the incentive cannot match the monetary return that would be realized by ignoring the voluntary environmental safeguards, as has often been the case, environmental degradation continues unabated.

April 24: Bush says "when you own property [e.g., his Texas ranch], every day is Earth Day." *Too bad it hasn't dawned on him that the majority of the American public feels the same way about THEIR federal lands.*

May 19: The president is playing a shell game to create a facade of environmental sensibility that obscures his anti-environmental bent. A case in point is his decision to designate a large section of the Chugach National Forest as wilderness while expanding commercial exploitation in the Tongass National Forest. This is a specious tradeoff. The protected area in the Chugach is a remote, primarily rocky region that few would venture into on any account, while the Tongass acreage in question is part of the nation's last remaining biologically diverse temperate rainforest.

May 23: Bush opposes an investigation into his pre-9/11 handling of terrorism out of fear it will expose incompetence, not malevolence.

May 30: It's comforting to know that for the right price, President Bush can be persuaded to do the right thing. Florida's environment is a case in point. To help his brother's reelection chances, the president has *suddenly* and *briefly* turned green and declared the state's off-shore waters out of bounds for energy development. He also employed a clever caveat to dash any hope that his Florida move represented an environmental epiphany. Bush said his decision reflected respect for grassroots sentiments—and guess what? In virtually every other wild place around the country, the only local voices he seems to hear belong to those shouting "drill, drill, drill!"

June 19: Bush recruits fringe groups with ostensibly legitimate environmental credentials to support his weakening of environmental laws.

June 27: The corporate scandals provide an opening for environmentalists to attack Bush, who has as a cornerstone of his environmental policy a heavy reliance on companies complying voluntarily with anti-pollution regulation. He also is a product of the corporate culture that is under fire, and from which most of his best buddies and advisors hail.

July 26: Bush is unlikely to attend the Johannesburg summit. He knows many nations will make him a target of criticism for not sufficiently sharing the United States' wealth and expertise. Anticipating these allegations, he is either unwilling or unable to defend his policies. Moreover, I don't think Bush really grasps the ramifications of his stonewalling. Despite his international exposure simply by virtue of being president, Bush's universe remains essentially bounded by Crawford, Texas; Kennebunkport, Maine; Washington, D.C.; and undoubtedly in the not-too-distant future, Baghdad. *Such a limitation puts him—and by extension, us—at a disadvantage.*

August 20: Bush wants other countries to cooperate in his war on terrorism, but he thumbs his nose at many of their concerns. *Speak of a double standard!*
 . . . Instead of clearing the brush on his Crawford, Texas, ranch, Bush should be clearing the air in Johannesburg, South Africa.

August 21: Bush is devoting virtually all his work sessions during his vacation to plotting for the conduct of war, while spending no time at all planning for a durable peace.

August 23: Bush's penchant for secrecy, acquired in corporate boardrooms, spills over into environmental policy. A prime example: keeping administration announcements to allow mining on BLM lands out of the Federal Register, which would alert environmentalists who might possibly react by mounting legal challenges.

August 24: Bush: "We will guard against excessive red tape and endless litigation that stand in the way of *sensible* forest management decisions." *Pul—lease!*

August 25: U.S. officials say Bush is too busy with national security concerns and the economy to attend the Johannesburg Environmental Summit. Those excuses don't square with photos of the president playing golf, clearing brush on his ranch, and speaking at fundraisers.

September 17: I wish Bush would quit describing us as the "greatest" or "best" nation in the world. "Great" and "good" would suffice. To constantly use superlatives implies that we are better than everyone else, and is bound to offend many people in other countries.

October 7: Bush has little interest in environmental concerns, not only because of his preoccupation with terrorism, but because he doesn't connect the dots between environmental quality and the human condition.

November 5: Bush's vision of the future consists of a world where oil derricks dot the wilderness landscape, urban centers battle vehicle and factory emission-induced smog, and population explodes and maternal mortality rates skyrocket globally, as reproductive health services and sex education are severely circumscribed because of "moral" objections to contraception and abortion. Nature is in full retreat as developers are given virtual carte blanche. Instead of living in harmony with the natural world, humanity treats it as an adversary that must be conquered by technologically driven solutions in the form of barricades, filters, underground shelters, and synthetic materials and products. The idea is to improve upon what we have destroyed. Along those lines, preservation of natural resources is not terribly important compared to economic development, which produces technological innovation that is supposed to compensate for any damage inflicted on the earth's natural systems. It's a world where America knows best because we *are* the best, and thus at the end of the day should call the shots for other nations as well as ourselves.

November 22: No sooner has the ink dried on the midterm election that gave the GOP control of both houses of Congress than the Bush White House has issued a flurry of anti-environmental administrative rulings ranging from rolling back the Clean Air Act and rescinding the ban on snowmobiles in Yellowstone National Park to authorizing new drilling for natural gas within the confines of Padre Island National Seashore. Is this the start of a pattern that will prevail for the next two years? The timing of the announcements, if nothing else, is ominous.

November 25: One gets sick of Bush's sanctimonious pap. "If our values (freedom, parental devotion) are good enough for our people, they ought to be good enough for others." There seems to be an assumption that these values are more deeply embedded in us than in some other nationalities.

November 26: How long are the American people going to let concern over national security distract them from reacting to Bush's corporate cronyism, obsessive secrecy, and efforts to roll back environmental protection and civil rights? When will they realize that public health and traditional conservation values are being seriously undercut?

December 5: Just as Democrats underestimated Bush's political acumen, so does Bush—like his father—appear to underestimate the political potency of environmental issues.

December 9: From a political standpoint, 9/11 is the best thing that ever happened to George W. Bush. Otherwise his administration would be in the metaphorical "toilet."

December 26: I wish Bush would quit threatening various nations around the world and ditch the menacing rhetoric for quiet, if firm, diplomacy. The United States has never been known as a bully, but that is rapidly changing, and our national security will deteriorate accordingly.

2003: . . . And On

January 2: Bush took a four-mile nature walk on his 1500-acre ranch today while waxing lyrical about his land's topography. *Speak of the blind leading the blind.*

January 5: Bush's obsession with differentiating himself from Clinton's policies regardless of the merits has frequently resulted in embarrassments, and sometimes costly national blunders.

January 12: Bush scornfully calls environmentalists "green, green lima beans" [1]. I don't mind the epithet, seeing it's coming from an arguably "rotten tomato".
 . . . The President just doesn't get it regarding wilderness preservation. He has allowed his bias in favor of the private sector to be transformed into bias against the public sector, an unfortunate turn of events when you consider public control of natural resources is less susceptible than the private sector to corrupting business influences.

January 13: Bush's frat-boy mentality has made him a loose cannon whose incendiary rhetoric spouting a pre-emptive strike philosophy has ignited unprecedented anti-Americanism throughout the Middle East, Asia, and even Europe. It also threatens to trigger several military conflicts that could put many Americans in harm's way.

January 14: It galls me when the Far Right praises Bush for having the guts to order our troops into Iraq. The true guts are displayed by our soldiers, because if Bush is wrong, all he does is go back to Crawford, Texas, while they go back home in body bags.

January 21: *New York Times*: "The Administration is playing to a base of hard-core ideologues who have this bizarre notion that conservation and environmental protection are left-wing plots against the American way of life."—Jim Di Peso, policy director for Republicans for Environmental Protection. *Jim, you got that right!*

January 23: "George W. Bush ranks with Teddy Roosevelt as an environmentalist"—Karl Rove. *Dream on!*

January 26: Stow the bluster, Mr. President, and stop telling everyone we are the greatest nation in the world. If you feel that way, let our actions make your case.

February 9: A religious zealot can be very dangerous, and we are being led by one. Spare me the platitudes, Mr. Bush, and defer to others more qualified to preach about morality.

February 10: Bush believes God chose him to free us from the scourge of terrorism and liberate Iraq. *Wrong! A minority of American voters and a majority on the Supreme Court did.*

February 13: If your defense of Bush consists of blaming Clinton, you are suppressing your doubts about our current leader. Remember, Clinton is not president any more.

February 15: Military success in Afghanistan emboldened Bush to shoot off his mouth and alienate most of our allies with cocky declarations about the superiority of the United States. *Incidentally, it was the Northern Alliance that did the bulk of the fighting on the ground and suffered the brunt of the casualties in the ouster of the Taliban regime.*

February 17: Bush is fueling fear throughout the nation. Is it a calculated maneuver to cover his flanks, or is he also a victim?

March 16: I can't decide which word best describes Bush's environmental persona—ignorance or insensitivity. Come to think of it, both are appropriate.

March 21: It's distressing to observe the detached manner in which the news media report on the bombing and killing of the Iraqi people (as opposed to American casualties). There is something surreal about this dispassionate disconnect to the loss of life, especially when it is downplayed and dismissed as "collateral damage."

March 30: Bush's statement during an election debate with Al Gore: "If we are an arrogant nation, they [the world] will resent us. If we are a humble nation but strong, they will welcome us." *Foresight is NOT forewarned.*

April 14: Emboldened by his lofty poll numbers due to the prosecution of the war, Bush no longer conceals his abject contempt for the environmental movement, and has aggressively accelerated deployment of his anti-environmental agenda.

April 17: So far, the biggest political joke of the early 21st Century is President Bush's fervent pledge "to be a uniter, not a divider."

April 24: As far as Bush's intellect is concerned, anyone who can get a "C" at Yale can't be all that dumb.

April 28: It is unseemly for our non-combatant "Commander in Chief" to celebrate "victory" and pose as a heroic military leader in an operation that cost the lives of American soldiers and Iraqi civilians, and is not over by a long shot.

April 30: Viewing abortion as a legitimate last resort is not a pro-abortion stance.

May 1: President Bush may have high approval ratings in the United States, but have Americans ever stopped to consider why, outside of our nation, he is probably the most reviled leader on the world stage, even in countries we count as friends and allies?

May 29: My stomach turns when I see Bush strut and swagger in front of our troops on the deck of a carrier as though he led the charge into an enemy foxhole.

June 13: Bush has to cling to his core beliefs to keep in place a thin veneer of calm and steady resolve that covers a fragile, tightly wound psyche.

June 23: Bush must be challenged by contrasting his dark image of the world with an optimistic one.

June 30: One wonders if Bush is even dimly aware of the philosophical anthropocentric underpinnings of his ideology's environmental philosophy—namely, that nature has no intrinsic value of its own and should be totally subservient to humanity.

July 2: Safely ensconced in the White House as he prepares to hopscotch around the country raising millions of dollars for the GOP, tough guy Bush dares Iraqi guerillas to "bring it on" against our troops. *What bravado!*

July 16: When are the American people going to wake up and realize they've been had?

July 18: *The understatement of the decade:* George Bush says "he doesn't do nuance."

August 4: Bush's penchant for secrecy reflects either fear of disclosure or a paternalistic, dismissive attitude toward the public's capacity to participate in the shaping of its own destiny. *Why not a combination of the two?*

August 10: I know it is standard practice for presidential aspirants to engage in campaign fundraising during the summer preceding an election year. But there is something obscene about Bush raising millions of dollars at barbecues for fat cats while the troops he sent to Iraq are suffering fatalities on a daily basis.

August 15: Bush is actually smarter than I thought. I just ran across his quote: "You can fool some of the people all of the time, and these are the ones you need to concentrate on." *That just about sums up his political strategy.*

August 16: Bush uses his weekly radio address to declare that "litigation often delays" fire prevention projects in national forests. The statistics say otherwise. His "healthy forest" initiative is a thinly veiled "license to cut" for the timber industry.

August 25: Bush declares "there is too much confrontation when it comes to environmental policy. There is too much zero-sum thinking. What we need is cooperation, not confrontation." *Why doesn't he practice what he preaches?*

August 26: At a Redmond, Oregon, political stop, Bush declares "You know you are in pretty good country when you see a lot of cowboy hats in the crowd." *Are you beginning to understand why, outside the United States, Bush is among the most despised of national leaders?*

August 27: Bush boasts of cutting fast-growing cedar trees on his ranch to save old-growth timber, but when it comes to protecting old growth on lands belonging to all Americans, he is nowhere to be seen.

August 30: Bush fans terrorist-inspired national paranoia to paralyze the public into docilely accepting what would otherwise be unpopular domestic and foreign policies.

September 14: We are custodians, not masters of our environment, a distinction that I don't think Bush clearly grasps.
 ... According to the Washington Post Magazine (9/7/03), the French tend to view Bush as a "squinty-eyed bully, a cowboy given to shooting first and asking questions later." He is also regarded as "something of a yokel—uncultured, unschooled, inarticulate, anti-intellectual, dangerously shallow—elected and supported by a populace too fearful of terrorist attacks, an electorate that values too much the blunt, common touch, and too little the more complex virtues of a renaissance man." Although *the French are pretty harsh, I would still give them an "A Minus" for their analysis.*

October 6: Rush Limbaugh is one of the most polarizing figures in the nation, yet President Bush publicly lauded this demagogic conservative radio talk show host as a "great American" upon learning of his drug addiction. *So much for Bush's vow to be a "uniter" rather than a "divider."*

October 10: Bush: "The principles I bring to office will not change with time or the polls." *Translation: I'm not budging from being a right-wing ideologue.*

October 27: Senator James Inhofe: "Bush has the best environmental record of any president in history." *The only question that comes to mind is "What is Inhofe smoking?"*

December 3: Bush lets down his guard momentarily and reveals his true environmental philosophy while signing his "healthy forest" initiative. He declares: "We don't want our intentions bogged down by regulations. We want to get moving." *Industry's extraction of natural resources on public lands shouldn't be hindered by pesky environmental constraints, right, George?*

December 10: Doug Bandow of the *American Conservative*: "Bush is remarkably incurious and ill-read. Good instincts can carry even a gifted politician so far. A lack of knowledge leaves Bush vulnerable to simplistic remedies to complex problems, especially when it comes to turning America into the globe's governess." *Right on, Doug!*

December 11: Bush snickered when told of accusations that the United States was violating international law by prohibiting countries that didn't join our war effort from bidding on Iraqi reconstruction contracts. Displaying mock concern, he quipped, "I don't know what you are talking about. International law? I better consult my lawyer!" *This sophomoric outburst is symptomatic of why Bush is not fit to be president.*

December 26: Bush's constant use of the term "homeland" conjures up the same disturbing ultra-nationalistic imagery as the Nazis' repetitive invocation of "the fatherland" in their official pronouncements.

December 28: Bush doesn't seem to grasp that we are battling a movement rather than simply bands of terrorists. When will it dawn on him that ultimately the only way to achieve enduring victory against terrorism is not through the gun, but by overcoming the widespread belief that we are waging a war against Islam? He can do that by demonstrating convincingly our sympathies towards Arab populations who seek to extricate themselves from dictatorial rule. Other strategies would involve extending as much assistance as possible to combat Middle East poverty and pursuing in earnest a resolution to the Israeli-Palestinian conflict.

December 30: Bush and his ideological cronies believe the American public cannot enjoy a healthy life and environment without economic prosperity. That leads them to the conclusion that business' concerns must come first. *I don't think the majority of Americans subscribe to that order of priorities. I certainly don't.*

2004: Coda

January 1: It is one thing to create a strong national defense infrastructure; it's quite another to stoke people's paranoia and attempt to turn the country into a giant armed camp at the expense of vital domestic needs.

. . . The Mad Cow disease outbreak is a perfect illustration of the Bush Administration placing more emphasis on catering to the quest for profit than maximizing protection of public health. Bush's decision to block regulation of the livestock industry was only reversed *after* Mad Cow was detected. *It often takes a calamity for Bush to extend the public the benefit of the doubt.*

January 2: Bush went recreational quail-hunting the other day. If I were sending young Americans into battle, and several of them were being gunned down virtually daily, I wouldn't have much stomach for firing a weapon at any living creature. *But then again, that's just me.*

January 25: Bush's worst enemies call him "ruthless, corrupt, untrustworthy, close-minded, authoritative, inarticulate, intellectually challenged, programmed, cynical, dishonest, violent, a draft dodger, and a religious fanatic who believes he speaks to and for God." [2] *Some, but not all, of this is overstated. It's your move to pick and choose.*

February 1: Bush believes he is a standard-bearer of democracy, morality, and a better way of life in general for the rest of the world. As such, whatever means he must use to spread this way of life across the planet is legitimate. *The end apparently justifies the means, and where have we heard that rationale before?*

February 2: Bush is using homeland security and the war on terrorism as budgetary excuses for shortchanging environmental protection and a myriad of other domestic programs. It is his attempt to demonstrate to hard-core conservative supporters that he is serious about reducing the size of the federal government, which in reality is expanding. *His military expenditures far exceed the domestic spending cuts. All Bush is doing is downsizing the federal government's role in helping Americans improve their daily lot.*

February 8: *Response to Bush's hour-long appearance on "Meet the Press":* The president genuinely believes he is doing what is right at home and abroad. Because his fervent conviction is the result of an ideologically rigid, myopic, and overly simplistic view of the world, it is unconvincing to many people at home, and to even more overseas.

February 19: President Bush says he is "troubled" by San Francisco allowing unsanctioned gay marriages to take place. *Gee, Mr. President, I hope you are not losing too much sleep over this since you need all the energy at your command to deal with terrorism, the Iraq morass, environmental degradation, and national job loss. Besides, isn't marriage better than the sin of promiscuity with which the gay lifestyle has been so closely identified in the past?*

March 3: Bush is, as he insists, "a uniter, not a divider," but the unintended application of the phrase relates to his political enemies. He has so alienated them that they are unified as never before in their determination to oust him from office.

March 19: Bush: "There is no neutral ground between civilization and terrorism." *Agreed!* Then Bush adds: "There is no neutral ground between good and evil." *But what is "evil" for Bush? Many people in Islamic countries disapprove of terrorists' methods but share their hatred of the United States. Are they evil?*

March 25: Right-wing radio talk show demagogues can say whatever they want to say, however outrageous, because their most ardent listeners will hear only what they want to hear.

April 14: EPA Administrator Michael Leavitt, as smooth an orator as ever graced the National Press Club, told an audience that "there are tensions between environmental aspirations and economic desires" and that it was his job to go for the best environmental result while "maintaining our national economic competitiveness." That is the Bush philosophy in a nutshell. The question is, what happens if there is no wiggle room left to strike a balance, at least in the short term, and a choice must be made between American companies and public health? *I fear national economic competitiveness would get highest priority with the Bushies, even though any short-term economic disruption caused by curbing illness and environmental deterioration would disappear over the long run, leaving society better off fiscally as well as health-wise.*

. . . President George W. Bush has been a "sidelines" sort of guy on the athletic field and the field of battle. His father wasn't.

April 15: Representative Eric Cantor of Virginia, the only Jewish Republican in the House of Representatives, declared that President Bush's strong support of Israel wasn't going to win over "very liberal American Jews, who are not going to be able to put aside their environmental or abortion politics." *Isn't this an expression of Bush followers' self-doubt about their environmental image, despite brave talk to the contrary?*

May 5: In an interview on Arabic television regarding American soldiers' abuse of Iraqi prisoners, President Bush said a military investigation was under way and declared, "We have nothing to hide. We believe in transparency. We are a free society." *Transparency? Is he delusional, or is he a liar among liars, which would be quite a feat in a town notorious for producing the best in the business? Maybe there will be transparency after the fact in the Iraqi case, but the window looking in on the formulation of energy and environmental policy in the Bush Administration has been opaque from day one.*

Chapter Two

Setting The Stage

Dual Personalities

Although George H.W. Bush was weaned in a business climate shaped by the Texas oil industry, it looked for a while as though he would completely transcend his roots when he entered public life. He became something of an internationalist, honing his legislative skills as a Texas congressman and his diplomatic expertise as our representative to the United Nations and ambassador to China.

By contrast, George W. Bush gloried in wearing his Lone Star State provincialism on his sleeve. You would never know from his down-home drawl and garbled syntax that he spent much of his youth in New England, attending a fancy prep school and two elite universities. He assiduously cultivated a "good old boy" image when transplanted to Texas, and carried the persona straight into the White House. The image quickly wore thin in Washington circles and was even more scorned abroad, where it reinforced the stereotypical view of Bush as an arrogant, insular "cowboy."

Closet Environmentalist?

During the elder Bush's years in Congress, he compiled a moderate environmental record. In the late '60s, he was one of the leading proponents of family planning assistance to the developing world. Success in curbing population growth could "well determine whether we successfully solve the other great questions of peace, prosperity, and individual rights that face the world," Bush declared in a 1973 speech.

As vice president under Ronald Reagan, George H.W. Bush headed a presidential commission that recommended increased federal purchase of parkland. And because Bush took pains to keep a low profile in a virulently anti-environmental administration, his benign stance tempered "green" opposition when he sought to succeed his boss in the White House. Indeed, Bush's campaign pledge that he would be the "environmental president" allayed the fears of many who were weary from eight years of unrelenting combat against Reagan's deregulation crusade. Many people voted for Bush utterly convinced he was a "closet environmentalist" who had no choice as vice president but to comport himself as a "good soldier" during the harsh Reagan years.

Early in his term, George H.W. Bush seemed on course to make amends for his predecessor's clumsy, oft destructive handling of environmental policy. He pushed through some stringent amendments to the Clean Air Act and appointed William Reilly, president of the World Wildlife Fund and a conservation activist, to head the Environmental Protection Agency.

Short-Lived Honeymoon

The environmental honeymoon was short-lived. Bush Senior soon became disenchanted with the degree of enthusiasm and support exhibited by the mainstream environmental community in response to his overtures. By his third year, a disillusioned George H.W. was well on his way to reverting to Ronald Reagan's mode of environmental governance. As the 1992 election approached, Bush felt increasingly compelled to consolidate conservative support and began attacking environmental policies that he once promoted. He disparagingly referred to Al Gore as "Mr. Ozone Man," and assigned Vice President Dan Quayle to revive the environmental deregulation crusade in a much more aggressive manner than Bush had displayed under Reagan.

> The elder Bush backed away from increasing federal funding for renewable energy and abandoned any semblance of an aggressive stance towards global warming. He curtailed his activist approach toward population issues and frequently sought to undermine preservation of public land by pushing for greater access for extractive industries.

Bush's "green" vestiges evaporated quickly, and he set about establishing a pattern of official conduct that his son dutifully emulated eight years later. The elder Bush backed away from increasing federal funding for renewable energy and abandoned any semblance of an aggressive stance towards global warming. He curtailed his activist approach toward population issues and frequently sought to undermine preservation of public land by pushing for greater access for extractive industries. Bush reverted with vengeance to his original philosophy that what was best for industry's bottom line was invariably best for the environment and the country. Voluntary compliance increasingly supplanted environmental regulation as the elder Bush's strategy of choice for gaining corporate America's cooperation.

In the end, wrote historian Kevin Phillips, Bush Senior could not rid himself of a "vocational hauteur bred from a lifetime of handling money for rich people." It was this attitude, Phillips added, that led the Bushes to view "the economic top one percent of Americans as the ones who count." [3]

The Isolationist Tinge

In regard to global environmental problems, Bush Senior's internationalist image took an isolationist hit. He unleashed his own "bully boys," foreign policy advisors, who had no qualms about throwing the United States' weight around in heavy-handed fashion. When other world leaders chose to attend the historic 1992 Earth Summit in Rio de Janeiro, Bush had to be persuaded to make a token appearance to avoid embarrassment. His ultimately meaningless gesture was consistent with the United States' role at the gathering as the odd man out, a portent of things to come. A decade later, Bush Junior passed up the Johannesburg, South Africa, follow-up to the Rio environmental summit altogether. He opted to chop wood on his Texas ranch and host barbecues for campaign contributors while other heads of state gathered to seek consensus on tackling the most profound issues of our time. Obviously, cross-pollination with other world leaders on global environmental problems was not a matter of importance to Bush Junior.

During his campaign, Bush Senior had called for an environmental summit to shape a framework for international cooperation. But when 77 out of 80 participating countries at a 1989 United Nations conference endorsed a 20 percent reduction in carbon dioxide emissions by 2000, Bush labeled those nations "extremists."

In April 1990, Bush Senior invited 17 countries to Washington for a conference on global warming. But he angered the participants when none were allowed to speak at any of the plenary sessions that were essentially used as platforms to advocate his government's minority dissenting views. While temporizing over global warming in the international arena, the elder Bush proceeded to low-ball funding for energy conservation and mass transit on the domestic front.

Painless War

During the 1991 Gulf War, the only sacrifice Bush asked of the American people was to voluntarily conserve energy. He appealed to the public to refrain from exceeding speed limits, keep their vehicles' tires at the proper pressure, and engage in car pooling. Let's just say that in response, there wasn't a mad dash in this country to revamp driving habits.

By refusing to push for a gasoline tax increase, raise energy conservation requirements for federal facilities, propose mandatory auto inspection, and provide tax incentives for renewable energy development, Bush in effect allowed himself to be led by those who had elected him to lead. His refusal to call for any serious national sacrifice in time of war presaged the policies of his son. George W.'s message to the nation at the onset of the overt war on terrorism was to urge Americans to bolster the economy by descending on shopping malls. He made no mention that the prices they encountered in the stores would be a pittance compared to the price that many of our young men and women in the military would have to pay in Iraq.

I've Got A Secret

Where Bush really set the tone for his son's administration was in a penchant for secrecy. To the Bush clan, it's not a "government of, by and for the people." It's a government that, to be effective, must operate behind the backs of people. This exclusionary mentality appears to be a carryover from the corporate boardroom, where chief executives normally have the prerogative to pick and choose when they will answer to their subordinates.

Bush Senior was conspicuously unenthused when heads of state at the 1992 Earth Summit in Rio de Janeiro officially agreed to the following declaration: "Individuals should have appropriate access to information concerning the environment that is held by public authorities, including information on hazardous materials and activities in their communities, and the opportunity to participate in decision-making processes . . . ."

The response of Michael Boskin, chairman of the president's Council of Economic Advisors, was to defend Bush's proposed restrictions on public participation as "efforts to remove unnecessary obstacles to business expansion and job creation." No surprise, then, that Bush Senior sought to narrow citizens' standing to sue the government. The president also lobbied for industry to obtain the right to change the terms of emission permits without public review.

The Legacy

While Bush Senior expanded the national park and wildlife refuge systems and was responsible for several other environmental good works early in his term, he was never able to transform his scattered successes into a consistent, cohesive policy. Plagued by indecision, he vacillated wildly, undercutting achievements with backsliding. It was a classic case of trying to please everybody and ending up pleasing no one. Although he often stated that the nation could have both robust economic growth and a healthy environment, he could never in practice bring himself to view those two elemental objectives as inherently compatible. He was always seeking to "balance" them, an exercise that by its very connotation signified an adversarial relationship and simply became a euphemism for a giveaway to industry. Bush Senior succumbed to the political pressure from the Republican Far Right, which opposed virtually any governmental intervention in the workings of the marketplace, even on behalf of environmental protection. He ended up allowing roads to be built in previously roadless areas under consideration for wilderness designation, and spent proportionately far less on the Environmental Protection Agency's budget than was allocated a decade earlier. [4]

The sequel 10 years later possessed much greater clarity. A more ideological son unencumbered by the ambivalence that plagued his father had no difficulty practicing uninterrupted corporate favoritism right from the start.

Chapter Three

Texas True And Blue

Only So Far

In 1999, Texas Governor George W. Bush approved a statewide renewable electricity standard that required utilities to provide a set percentage of their output in non-polluting renewable energy. That action helped generate $1 billion for investments in wind power and made Texas a world leader in the production of that form of energy.

That's the good news. The bad news is that Bush was backed into taking the action. It required the intervention of a major campaign contributor who owned a clean energy company and the threat of U.S. Environmental Protection Agency sanctions due to state air pollution violations to persuade Bush to sign off on the initiative. [5]

Further indication that Bush was a passive participant rather than a driving force behind wind power's ascendancy in Texas came when he assumed the presidency. It was then that he opposed enactment of a mandate that renewable energy comprise 10 percent of utilities' sales throughout the *entire nation* by the year 2020. Texas' success evidently wasn't sufficient to trump his ideological bias in favor of letting the marketplace, rather than government leadership, sort things out.

The Lone Star State was not as fortunate with air pollution as it was with wind power under Bush. Texas led the nation in factory emissions of toxic and ozone-producing chemicals, including carcinogenic compounds particularly damaging to children's neurological development. While Bush held sway in Austin, 64 percent of Texans resided in areas that failed to meet the EPA's clean air standards. The public interest organization Environmental Defense Fund conducted a survey of 21 air quality indicators and found that every one of them deteriorated during Governor Bush's term in office. It was determined that Texas refineries emitted three times more pollution than their counterparts in other states. Hence, Houston's 1999 usurpation of Los Angeles as the metropolis with the worst air quality in the country should have come as no surprise. Neither should have the disclosure that not a single piece of clean air legislation was proposed by Governor Bush, despite Texas' dubious distinction as the most polluted state in the nation.

It was not only environmental indifference on the governor's part that played havoc with Texas. Bush exempted 760 Texas-based oil, gas, and chemical companies from mandatory emission reductions and instead asked for *voluntary* compliance. The governor's plea for cooperation fell on the deaf ears of his corporate cronies. Less than 10 percent heeded his call, which is why miasmic air lingers to this day over Texas' industrial urban centers.

On the campaign trail, Bush promised to provide federal financial resources to help states and municipalities set aside conservation areas. The money would also be used to furnish private property owners with tax incentives to preserve land and wildlife. These were good ideas as long as they didn't detract from the preservation and upkeep of national parks and other federal wilderness holdings. But given Bush's philosophical underpinnings, one feared the worst.

There is no question it is desirable to get the support of local communities for any increase in wilderness protection of neighboring federal land. Bush behaved, however, as though he was oblivious that proximity conferred no special proprietary privilege regarding public land. Federal acreage belongs to all

Americans, and thus the will of the public (reflected in the management decisions of our representative government), not the wishes of a particular segment of the population, is what should determine the ultimate status of the land.

Bush's gubernatorial record reflects this ideological anti-federal prejudice. In Austin, he boasted that less than three percent of Texas was owned by Washington while proudly asserting that 90,000 new acres were added to state parks during his tenure. What Bush failed to mention was that he had no role in the acquisition of virtually all the acreage. Conservation organizations deeded the land to Texas as a gift. If the truth be known, Bush fiscally starved the state's land acquisition agency, with Texas ranking 49th in per-capita spending on the purchase and maintenance of parkland. The state thus couldn't have added much in the way of further parkland to the 90,000 acres even if it wanted to. One final negative: Texas' undersized state park system had a $186 million maintenance backlog during Bush's tenure.

Any way you look at it, conservation was not Governor Bush's forte.

The Platform Tell All

The 2000 Republican presidential platform reflected George W. Bush's views, and as such was cause for concern. Unfortunately, the public rarely pays attention to pre-election party platforms, even if there is a good chance they are a portent of things to come.

In Bush's case, it was especially lamentable that the American people weren't more heedful of what was contained in the campaign screed. If they had, they would not have been so accepting of Bush's pledge that he would be a uniter rather than divider.

In the document, the GOP made no secret of its intention to increase reliance on voluntary corporate compliance in lieu of aggressive enforcement of environmental statutes. The platform was also quite transparent about placing regulatory enforcement responsibility to a higher degree than ever before in the hands of the states.

Little criticism was leveled against this delegation-of-responsibility approach, whose flaws were graphically illustrated by Bush's own record in Texas. His appeal for voluntary cooperation from corporate polluters produced few positive results, perhaps because his oil industry buddies sensed that their friend in the statehouse would have infinite patience with their stonewalling.

The GOP platform contained a pledge to make corporate America "parties with government rather than adversaries of it . . . Scare tactics and scapegoating [a.k.a. regulation] of legitimate economic interests undermine support for environmental causes." The federal government's "main role should be to provide market-based incentives" to clean up. *A promise to strictly enforce stringent environmental regulations was conspicuously absent.*

Bush's staunch faith in the states' fulfilling their environmental enforcement responsibilities with flying colors raised a red flag during his campaign then, as it does now.

A recent analysis of states' regulatory performance found a very spotty enforcement record vis-à-vis the Clean Water Act. It was clear that some state water quality agencies were under the thumb of industrial polluters in their jurisdiction. And water wasn't the only problem. The Washington-based Environmental Working Group's review of government data disclosed that one-third of the nation's major air polluters had not been inspected by the states between 1997 and 2000.

Although the platform contained no explicit pledge to sell off public lands to private interests, it was easy to see where Bush's sympathies resided. With his approval, Republicans asserted in the platform that "the world's worst cases of environmental degradation have occurred in places where most property is under governmental control." I assume the Grand Canyon, Yosemite, and hundreds of other national parks and refuges are exceptions to the rule, or would they be if Bush had free rein?

In His Words

It was bad enough that the GOP platform was largely ignored during the campaign. The sin was compounded by lack of attentiveness to George W. Bush's own words, which gave plenty of clues as to what he would do if elected president.

While he declared a love for wilderness, he complained about public lands being commercially underutilized and vowed to initiate oil drilling on the pristine Arctic National Wildlife Refuge as well as to increase logging in our national forests.

Although Bush acknowledged that global warming was real, he contended that we didn't know enough yet to justify immediate remedial action to curb the phenomenon, and he has maintained that position throughout his presidency.

Bush warned that he would be receptive only to federal environmental regulations that were based on "sound science." It was a standard that always seemed conveniently non-existent when his administration was intent on delaying regulation of industry. The Bush Administration would argue that no regulation could be justified until conclusive proof was forthcoming. Here is the problem with that approach. Responsible government officials cannot afford to wait for conclusive evidence that science may never be able to provide in our highly complex world. They must act on the *weight of evidence,* or they risk responding with corrective measures only after people start dropping in the street. Sadly, it's a risk that hasn't seemed to resonate with President Bush.

Uniter Or Divider?

Bush repeatedly pledged that he would establish a new conciliatory tone in fractious Washington. But if he were truly sincere, he would have given assurances that he would not dilute stringent environmental regulations that enjoy widespread public support. Instead, there were hazy allusions to repressive rules that needed to be rolled back, with the lack of specificity raising the specter of a new round of bitter political infighting. It was obvious to the politically astute that Bush's indebtedness to his conservative corporate base would discourage his promotion of many environmental initiatives with broad bipartisan appeal.

Candidate Bush was also deliberately vague about his core supporters' "hot button" issues. He declared his opposition to abortion but emphasized his support of the procedure in cases of rape, incest, or danger to a mother's life. He denied he was in the hip pocket of the National Rifle Association, and crisscrossed the country delivering populist speeches in which he promised to make the focus of his administration the improvement of conditions for low-income Americans.

> All the while, the NRA, corporate America, and pro-life organizations that wanted a total ban on abortion were campaigning furiously for the Republican nominee. They were confident that when push came to shove in the White House, their priorities would prevail despite Bush's campaign rhetoric—and they were right

All the while, the NRA, corporate America, and pro-life organizations that wanted a total ban on abortion were campaigning furiously for the Republican nominee. They were confident that when push came to shove in the White House, their priorities would prevail despite Bush's campaign rhetoric—and they were right.

Prophetic

Midway through the campaign, 60 environmental leaders from 18 states became so disenchanted with Vice President Al Gore downplaying the "green" issues that he had so long championed that they formed an organization opposing his candidacy. Although they considered Bush a far inferior candidate, they figured his victory would unify and mobilize the national environmental movement far more than an indecisive Gore in the White House. Did they ever get that right! The problem is they got more than they bargained for. It apparently didn't occur to them that the Republicans might retain control of Congress—as they ultimately did—dramatically reducing opportunities to thwart any regressive Bush initiatives. They forgot that it was often only a presidential veto that foiled a Republican Congress' environmental rollback during Bill Clinton's second term.

Egomania, hubris, and naiveté clearly were in play for these renegade environmental leaders to acquiesce to a candidate who posited a number of times on the campaign trail that clean air and clean water could not be produced through legislation or litigation. According to Bush's wishful thinking, only by urging polluters to voluntarily cease and desist, and by offering tax incentives to reinforce the request, could pollution be ultimately brought under control.

To link one's destiny to this pipe dream for whatever reason was asking for trouble big time.

Chapter Four

The Early White House Days

Anything But Clinton

One of the first acts of the newly installed president was to slap an immediate freeze on most pending federal regulations, for no other reason than antipathy towards Bill Clinton. There was great reluctance to grant instant affirmation to *anything* that could somehow contribute favorably to the Democrat's legacy. The merits of the Clinton proposals were lost in a swirl of vindictiveness, even in regard to a rule tightening the permissible limit for arsenic in drinking water.

A Senate investigative report described how incoming EPA administrator Christine Todd Whitman was briefed by her assistants in support of the Clinton arsenic standard, only to be told by the White House to withdraw the rule on the flimsy excuse of insufficient scientific justification. Such reasoning simply didn't hold water, considering the new standard had undergone years of extensive study and public hearings as well as received the endorsement of the National Academy of Sciences. Only after nine months of stonewalling did the George W. Bush Administration bow to pressure from an outraged public and reinstate the Clinton rule. For many, however, Bush's initial stance raised doubts about the depth of his commitment to environmental reform.

How Dumb Can You Get?

Call it naiveté. Call it stupidity. Call it desperation. On the eve of "W" taking office, there were some who had hope that his environmental policies would be a pleasant surprise. After all, the country was split down the middle politically, as attested by the close election, and the majority of people at both ends of the ideological spectrum considered themselves environmentalists.

Those harboring optimism were doomed to disappointment. The political stars were simply not aligned. There was no way Bush could satisfy his hard-core conservative political base without vigorously pushing a controversial partisan agenda that involved favoring industrial expansion over environmental protection. His campaign pledge "to be a uniter rather than a divider" was pulp fiction. He had officially been in office less than 48 hours when he launched what environmentalists considered the most intensive presidential assault on environmental protection in American history. Ideology prevailed over pragmatism in the crafting of policies to combat ecological abuse.

Translucent Compassion

How the second President Bush initially handled environmental concerns provided early insights into just what he really meant by "compassionate conservatism," and the revelations were disquieting.

On the campaign trail, he had defined "compassionate conservatism" as making sure government was not an intrusive presence and only exercised its authority when it could clearly make a positive difference in people's daily lives.

> ... the president's very first act in office was to reinstate the global gag rule.

Fair enough, but the president's very first act in office was to reinstate the global gag rule. His decree suspended federal aid to any foreign family planning program in which abortion counseling was provided, even if such a service were paid for with the organization's *own* funds. The result of this so-called "global gag rule" was to restrict funding for—and thus in many cases free access to—contraception in countries where abortion was legal (as it is here).

Hence, Bush's first official act had the effect of increasing the frequency of the very surgical procedure he was trying to discourage. Cutting off funds for family planning services on foreign shores led to unwanted pregnancies and an increase in abortions, not to mention setbacks in combating AIDS. The way was also paved for more women to die during pregnancy (including from botched abortions) because of a lack of access to artificial contraception. It was a very inauspicious debut for "'compassionate conservatism." Frankly, it came across more as an act of petty revenge against Clinton, who had rescinded the gag rule that was in place during the presidency of Bush's father. Even worse, it was indisputably ideological pandering to the political right, at the expense of maternal health care in developing countries.

When the Clean Air Act was approved decades ago, Congress in its wisdom decided that health alone should determine the regulatory standards for air quality. Cost was only to be factored into the equation in implementing the standards. That arrangement gave federal regulators leeway to provide an extra "margin of safety" for the more vulnerable elements in our society (e.g., the very young, the very old, and the very sick), even if the considerable added expense were for the benefit of a small percentage of the population.

Such an arrangement was an anathema to Bush, who moved quickly to put costs at least on a par with health concerns in regulatory formulation. Was "compassionate conservatism" showing its true colors?

Early Pluses And Minuses

After giving full vent in the first few months of his term to his abhorrence of Clinton, President George W. Bush seemed to relent to some extent. He opted to back (or at least not oppose) the previous administration's eleventh-hour regulations to combat lead pollution, require more energy-efficient washing machines, curb toxic pesticide use, and increase wetland protection. Bush proposed to end subsidized federal insurance to property owners who refused to relocate from vulnerable low-lying ground and thus were being compensated repeatedly for rebuilding their flood-damaged structures.

Was Bush experiencing an epiphany? Alas, there were some distinct caveats to this new president's ostensible change of heart. It turned out that Bush's enthusiasm tended to flag when it came to stringent enforcement of these environmental initiatives or vigorously defending them against legal challenges.

His proposal to eliminate federal flood insurance for people who recklessly put themselves in harm's way was commendable. But his preferred solution of relying on structural barriers, rather than conversion of flood plains into vegetative buffers, was only bound to exacerbate flood damage in the long run. He substantially cut federal programs to facilitate people's relocation from hazardous low-lying areas and help communities restore wetlands that could act as sponges along flood-prone waterways. Whatever his original intentions, when push came to shove, Bush could not say "no" to the real estate lobby.

James Watt In Skirts

All doubts about where George W. Bush stood on the environmental front were erased with his nomination of Colorado-based attorney Gale Norton to head the Interior Department. Norton was a protégé of Ronald Reagan's infamous Interior Secretary James Watt and assiduously worked to implement his regressive land-use policies. In line with her ultra-conservative and libertarian bent, Norton had spent a great deal of time demonizing the very department that Bush had asked her to lead. Norton had repeatedly called for Washington to cede much of its responsibility for managing federally owned lands to state and local officials, whom she believed invariably did a better job. And that bias has persisted at Interior, even if she has had to be subtle about integrating it into her decision-making.

Norton's propensity to side with private citizens, who maintained that their adjacent location to federal lands entitled them to a greater proprietary interest in that acreage than the rest of the country, was disturbing then. It is disturbing now. The truth is that the little old lady in a Brooklyn, New York, tenement has—by law—as much say about disposition of federally owned lands thousands of miles away as any individual living within a stone's throw of that territory.

Environmentalists worried that Norton would be a "James Watt in skirts." Their trepidation was well founded. Indeed, she was even more of a menace than first feared. Norton implemented her heavily ideological-driven policies in such a low-key way that she didn't attract the public furor that led to Watt's ouster from government.

Bush characterized environmentalists' opposition to his appointment of Norton as nothing more than overwrought rhetoric from "special interests." But the "special interests" he dismissed so cavalierly were not any single group with a narrow agenda. Public opinion polls showed repeatedly that the national environmental movement's agenda had the support of an overwhelming majority of Americans of *all* political stripes.

Thoughts Unspoken

What President George W. Bush had to say about the environment during his first address to Congress sounded just fine as far as it went. What wasn't mentioned in his budget speech were the sort of things the American people really needed to worry about. There was no reference to the many environmental programs that the president was planning to slash to help finance his proposed $1.6 billion tax cut.

While it was nice to hear Bush declare that he would ask Congress for the full appropriation reserved for public land acquisition (Land and Water Conservation Fund), he didn't elaborate on how the money would be distributed. If he had, you would have learned that the federal agencies that manage our national conservation areas were being drastically shortchanged in favor of the states. There would be insufficient funds to purchase all the necessary additions to our national parks and refuges and other federal areas set aside for conservation. Conversely, the states would have more money available than ever before to buy parkland. A fly in the ointment is that not all of the states have chosen in the past to allocate the extra money for that purpose.

The president indicated in his speech that renewable energy sources and conservation would play a major role, right alongside increased fossil fuel production, in the creation of his national energy policy. He did not elaborate on how he would achieve this objective, and with good reason. It was apparent in his proposed budget that he was actually trimming money from research and development of renewables and conservation (again, to support tax cuts). How did he plan to offset this loss? By replacing it with $1.2 billion in speculative revenues derived from drilling for unknown quantities of oil on the Arctic National Wildlife Refuge's coastal plain in northeastern Alaska.

It was a classic shell game. Bush did not have sufficient votes in Congress to hand over the unique ANWR wilderness to the oil industry, but that didn't stop him from proposing to fund environmentally friendly energy alternatives with money he didn't have, nor was likely to get.

President Bush was going to make reference to a new air pollution control strategy in his speech, but when some energy industry lobbyists objected to the thrust of his planned remarks, the words were deleted. Good environmental intentions rarely have had much staying power in the Bush Administration.

W's Brave New World

Over the past few decades, industry has rallied around the premise that it's fiscally counterproductive for government to focus on regulating "minor" risks, such as air pollution and pesticides. Authorities' attention should be directed at reducing "major" risks, such as driving, smoking, alcohol, and poor nutrition.

Industry is using a diversionary tactic in the hope of escaping liability. It is unlikely that Washington would ever play much of a role in regulating these "major" risks, which revolve around lifestyle choices that we are all free to make. Bush junior and his corporate cronies would surely bridle at any significant governmental attempt to regiment our daily life in this manner, as would we all.

"Minor" risks are different. They are created by industry's release of toxic pollutants into the environment. Although Bush's surrogates maintain that comparatively few people are threatened by these releases of "trace" pollutants, the risks to human health are certainly not trivial to the unlucky souls who fail to beat the (supposedly) favorable odds. Moreover, such risks are usually imposed upon the general population without their knowledge, and accordingly should be overseen by government regulators, because no one else can—or will—protect the unsuspecting public.

In fact, Bush would have the government refrain from regulating "major" risks, because they were questions of personal choice, and eschew policing "minor" environmental risks because of allegedly disproportionate costs. Corporate polluters would be off the hook every which way they looked!

Anyone But Clinton

Three months into his term, George W. was still was on the warpath against Clinton. Bush's deep-seated enmity towards his predecessor precluded letting "bygones be bygones." But in his eagerness to purge as much of the Clinton residue as possible, Bush rarely offered any convincing explanations for quashing his predecessor's environmental initiatives. Furthermore, he furnished no detailed description of how or when he was going to improve on the environmental measures he was rescinding, or why they warranted cancellation in the first place. He justified his actions only in vague general terms, citing "sound science" and "common sense." Without detailed amplification, these phrases are very subjective and arguably enable a politician to adopt whatever position he or she wants. That is just what happened. Bush used the euphemistic cover of the terminology to weaken the previous administration's environmental regulations and thereby take revenge against Clinton for defeating Bush Senior back in 1992.

Earth Day Déjà Vu

> *The first Earth Day under George W. Bush's reign bore an eerie resemblance to the celebration of that holiday under Ronald Reagan.*

The first Earth Day under George W. Bush's reign bore an eerie resemblance to the celebration of that holiday under Ronald Reagan. It was back in the 1980s that the country found itself in the clutches of Interior Secretary James Watt, a Reagan appointee who incited widespread public outrage by systematically seeking to unravel environmental safeguards at the behest of industry. Bush resurrected Watt's ill-conceived crusade with as much passion but far less fanfare, a stealth approach that created tremendous concern among environmentalists. They knew all too well that it's much more difficult to sidetrack subtle incremental environmental rollbacks than blatant ones.

Chapter Five

Arrogance And Unilateralism

Forget Nuance

"Truer words ere said in jest," which brings us to George W. Bush's quip, "You can fool some of the people some of the time, and those are the ones you want to concentrate on." Bush may not have felt his witticism was describing his own modus operandi, but I do.

One remark that unequivocally reflects Bush's persona is when he exclaimed, "I don't do nuance." I suppose we should be grateful for his candor, but it is an admission that this president lacks the cognitive tools to deal competently with the challenges to our nation over the long term in an ever more complex world.

The Ugly American

George W. Bush's distaste for nuance may have appealed to those Americans yearning for a simplistic approach to a turbulent international scene. But his single-mindedness won few friends abroad. Indeed, more often than not, Bush was regarded as arrogant, insular, uncultured, and just plain ignorant of the world outside the United States' borders. Some of this could be chalked up to jealousy of the United States' status as the world's sole superpower, but we occupied that role before Bush took office without the degree of enmity we now confront. At this juncture, Bush's "in your face" condescension towards the international community has been the leading anti-American catalyst in the 21st Century.

The outpouring of sympathy that the United States received throughout the world as a result of the 9/11 terrorist attack was squandered within a few months. Incredibly, Bush quickly became one of the most loathed leaders on the international scene, actually surpassing Saddam Hussein in some quarters.

> ... when Bush's charm fell flat, all pretenses of flexibility vanished, leaving a parochial, absolutist ideological stance for the entire world to see.

For most Europeans, no amount of Bush's Texas sweet talk could obscure the fact that he was ultimately telling, not asking, them what course to take. And when Bush's charm fell flat, all pretenses of flexibility vanished, leaving a parochial, absolutist ideological stance for the entire world to see.

The Not So Grand Obsession

There have been times when George W. Bush's obsession with ideological correctness has bordered on the irrational. A prime example was the White House becoming a de facto accomplice to firearms smugglers in its zeal to cater to one of its core constituencies. The Bush Administration successfully weakened international efforts to curb the illegal traffic of small arms and other weapons to appease the American gun lobby that balked at the idea of the slightest check on private citizens packing handguns.

In the environmental realm, the president objected to an internationally conceived plan that would have phased out government subsidies for fossil fuel projects in developing countries and used the money instead to fund renewable energy for a billion people without electricity. His opposition was based on a preference for letting the marketplace, rather than government subsidies (through international lending institutions or bilateral grants), bring about the expansion of renewables.

What he conveniently ignored was that market forces are often a fiction. Fossil fuels' far greater presence than renewable energy in developing nations is due in large part to huge government subsidies, not open competition. More than $115 billion in outside financial assistance was funneled to fossil fuel projects in the Third World between 1994 and 1999 alone.

Bush's opposition to government financing of renewable energy in the developing world angered our allies. They didn't like receiving ultimatums that it was Bush's way or no way. But that was hardly the extent of the damage. By emphasizing expansion of fossil fuel generation at the expense of solar, wind, and other clean renewable sources, Bush set the stage for further environmental degradation and the exacerbation of global warming. By denying millions of economically disadvantaged Third World citizens any access to renewable energy sources, and thus the only practical, immediate opportunity to introduce electricity into their lives, the president was also inviting additional public health problems and political instability.

Alieanation

The President's pledge to end the partisan bickering in the nation's capital and be a "uniter" instead of a "divider" has looked more like a cruel joke with each passing day of his term. His stubborn, ideologically motivated reluctance to move toward the political center actually exacerbated the divisions in Washington and even spurred a widening rift between moderates and conservatives in his own party.

As was previously noted, he alienated friend and foe alike overseas with a series of hard-line unilateral decisions in which our government essentially thumbed its proverbial nose at the rest of the world. He wasted no time rejecting proposed international agreements on global warming, germ warfare, and anti-ballistic missiles, despite their enjoying the support of virtually every country on earth.

Adopting this unilateral approach, without offering any specific alternatives, isolated us politically and economically from an ever-more-interdependent community of nations. Yet Bush was in denial about this separation. He fumed at the unilateral label, declaring that "unilateralists don't come around the table to listen to others . . . ."

Yes, they do, if their minds are already made up and their attentiveness to others' views is merely a pro forma gesture aimed at massaging delicate egos.

Prior to 9/11, Bush spiraled downward aimlessly in the polls on domestic environmental issues and just about everything else. [6] The terrorist attack brought focus to his presidency, and the country rallied behind him, with patriotic sentiments muffling public reservations about his environmental policies and other unpopular aspects of his agenda. Nonetheless, it didn't take Bush long to squander all that goodwill and revive the partisan divisiveness that erupted when his early actions contradicted his campaign promise to follow a centrist path.

President George W. Bush's advocacy of an "abstinence only" sex education curriculum for developing countries aligned us with such "progressive" nations as Libya and the Sudan. So much for his "forward looking" policies! Bush was siding with anachronistic societies in which no comprehensive reproductive health education exists. Communication is prohibited regarding contraceptives and their pivotal role in preventing unintended pregnancies and sexually transmitted diseases.

Bush's attempts to impose his Christian fundamentalist value system on nations with different cultural and social norms from our own regarding reproductive behavior were not only brazenly intrusive but an exercise in futility. His faith in abstinence's efficacy was also not borne out in this country. Recent studies have found that the rate of unintended pregnancies among teenagers instructed only in abstinence is roughly the same as that of kids provided with a full range of birth control options.

Even when Bush popularized "abstinence only" education as governor of Texas, his state was second in the nation in the rate of teen pregnancy and dead last in decline in teen birth rates among 15-to 17-year-olds. If it didn't work in the Lone Star State, why should it work in Africa? Bush's religious fervor evidently blinded him from contemplating the question, much less asking it.

To Be, Or Not To Be

President George W. Bush's decision not to attend the World Summit on Sustainable Development in South Africa at the end of August 2002 gave credence to his critics' claim that he was unwilling or unable to work with the international community in resolving global environmental problems.

> *The deterioration in our image was a shocking reversal for a country that was the driving force behind the creation of the global environmental movement.*

His absence heightened fears among friends and foes alike that he would steer the United States toward unilateralism, isolationism, and ultimately "rogue nation" status vis-à-vis the environment. The deterioration in our image was a shocking reversal for a country that was the driving force behind the creation of the global environmental movement. Bush's boycott of Johannesburg was even more appalling given the slowdown of the international campaign to upgrade conditions for the more than one billion people living on less than a dollar a day and without safe drinking water.

Image Making

Given the aforementioned backdrop, you've got to wonder about just how much George W. Bush was in touch with reality. He was actually perplexed at why so many foreigners voiced hatred of the United States. Rather than look in the mirror, he gazed out the window and saw the world's only super power humming industriously along, setting an admirable example for the rest of the world to follow. His solution was to create a propaganda office to repair our overseas image. The problem was that words alone wouldn't do the job.

Blinded by his "moral clarity," Bush didn't grasp that to turn things around, the United States had to cease being the odd man out. He didn't seem to understand that somehow we had to come to terms with international treaties—especially environmental ones—that virtually the entire community of nations, including all our major allies, supported.

If the treaties were good enough for everyone else, what made us so special? Until we could demonstrate convincingly that we were as good at taking others' concerns into consideration as we were at issuing ultimatums, we would be pariahs in many people's eyes, and that holds true to this day.

Bush also had to cease breaking the world down into a simplistic dichotomy of good and evil. Superimposing a black-and-white paradigm upon a gray reality stepped on too many toes. When Bush denounced certain nations as evil incarnate, many foreigners questioned what gave him the moral stature and authority to render sweeping ethical judgments on their countrymen. Bush's promulgation of verdicts particularly infuriated those who lived in nations where the United States had allied with repressive regimes.

It never crossed Bush's mind that his pontification was virtually interchangeable with rhetoric used by Saddam Hussein. Only the subject and the object of the verbs were different. Too bad, because there is no way Bush is going to improve his tarnished image (which has begun to rub off on his fellow Americans) without somehow divesting himself of his sanctimonious airs. Suggesting—sometimes not even subtly—that our lifestyle is superior to others is also a giant turnoff. There are those who think Bush's ultimate arrogance was to decide, without any consultation, that our nation should seek to remake Iraq in our own image. [7]

Poison Pen Letter

An August 2002 letter of support to George W. Bush from his conservative followers illustrated just how regressive the president's environmental policies are. Leaders of a number of well-known conservative organizations congratulated Bush for skipping the Johannesburg environmental summit that they characterized as a global media stage for "irresponsible and destructive elements."

And, pray tell, who were these "elements"? They were the heads of state and official delegations from more than 100 nations, along with scientists, trade unions, indigent people's representatives, grassroots environmental organizations, and many prominent private individuals from all walks of life. In short, the conservative leaders were marginalizing a cross-section of humanity, not exactly an ideal formula for building a mainstream political movement.

The conservative letter writers praised Bush for spurning a forum devoted to "various anti-freedom, anti-people, anti-globalization, and anti-Western agendas." Just what agendas were they talking about? The scheduled program for the environmental summit involved firming up international cooperation to reduce poverty, halt the spread of AIDS, establish universal primary education, and provide clean drinking water and adequate sanitation facilities for a billion people. Sure sounds subversive, doesn't it?

Bush never publicly commented on the letter, but he didn't disappoint its authors.

Above The Law

Environmental problems global in scope, the scattered locations of vital natural resources, and an increasingly integrated world economy have created a fundamental interdependency among all nations. Yet Bush had us thumb our collective nose at the international community, most notably through refusing to work toward implementation of an agreement to curb global warming. Eerily reminiscent of his father, who on a state visit to Canada was taunted by crowds for weakening environmental safeguards, Bush was jeered by thousands of street protestors for his environmental insensitivity on his first official visit to England.

The climate change pact was not the only widely endorsed treaty that we scorned. Just two countries failed to ratify the Rights of the Child Treaty—the United States and Somalia. We were among the few holdouts in approving pacts to ban discrimination against women, establish an international criminal court, and protect economic, social, and cultural rights. We dragged our feet in working toward the goal of the military arms reduction treaties that we had previously signed, including an agreement to draw down nuclear arsenals. And we were one of just 10 countries that refused to sign the treaty to halt the manufacture of antipersonnel mines.

We rejected arms reduction pacts because we considered provisions for verification inadequate. But our refusal smacked of a Catch-22. The kind of access Bush demanded to other countries' military installations was the kind of access he would not grant for their inspection of our own facilities.

Is it any wonder that we have won few international popularity contests of late?

The United States needs the cooperation of other countries, not only to conquer terrorism and prevent the spread of weapons of mass destruction, but to reverse environmental degradation global in reach. Powerful we are, omnipotent we are not!

Moral Clarity

The "moral clarity" attributed to George W. Bush's decision-making is in effect "ideological clarity" that is often anything but "moral." True ideologues are prone to cut ethical corners because of a fervent belief that the end justifies the means. *Where have we heard that maxim before?*

Driven by the single-minded pursuit of conservative ideological purity, the president and his advisors operate from the premise that, in the long run, the private sector will produce more benefits for society than governmental programs will. The "moral clarity" of this conviction exudes an ideological absolutism that can have negative effects.

Some examples: The president supported private companies taking over management of water distribution systems throughout the world. But such companies are in operation to make a buck, and sometimes this goal conflicts with providing water to all citizens regardless of their ability to pay.

Bush's distrust of the public sector led to delegation of responsibility for overseeing reconstruction of war-torn Iraq primarily to American companies, at the expense of the United Nations and humanitarian non-profit organizations' participation. The companies' dominant presence reinforced the image of our nation as an occupier rather than a liberator, an image severely complicating our reconstruction efforts.

Tomfoolery

George W. Bush's ideological rigidity resulted in some sophomoric diplomacy that would be hilarious if it weren't so irresponsibly muddleheaded. The president and his entourage may have resented France and Germany for opposing our Iraq policy. But to openly display pique and tacitly encourage public animosity against these two allies was politically moronic. Did it really make sense to incite boycotts against French fries and German beer?

What about the war on terror and our national security? Did it ever occur to Bush and his subordinates that we would be toast without the cooperation of France and Germany, two nations in the midst of the terrorist vortex?

Administration Francophobes, many of whom would snub the United Nations as well, seemed blind to the reality that major environmental problems transcend boundaries, and the cooperation of *every* country would be needed to reverse widespread ecological deterioration.

Bush didn't seem to recognize that international help was essential to phase out the massive use of highly toxic pesticides and other chemical pollutants whose residues drift halfway around the world and eventually land in our backyards.

Those who were so eager to purge France and Germany from collective memory also forgot that we needed markets for our products, and that "old Europe," as Defense Secretary Don Rumsfeld disparagingly labeled the two countries, was a key outlet for our merchandise. France and Germany could not easily be replaced as customers (or suppliers, for that matter), and any attempt to do so would grievously injure our economy. And do we really want to close ourselves off to a European cultural heritage that has enriched our civilization from day one? Consumed with ideological rancor and constricted by a provincial outlook, Bush has displayed an "it's my way or no way" attitude towards the countries of Europe. To the detriment of our national security, all too many Europeans have told Bush "no way."

Let The U.N. In

President George W. Bush's jingoistically motivated lack of enthusiasm for multilateral arrangements was reflected in his opposition to a United Nations presence in Iraq immediately after the war. No matter that it seemed to make sense, fiscally and otherwise, to recruit as many countries as possible to assist substantially in the reconstruction of the strife-torn country.

The error of his ways became increasingly evident in the ensuing months as the essentially isolated United States experienced an increasingly heavy burden in providing security and economic aid to Iraq. Miffed at Bush for being dismissed so perfunctorily months earlier, the U.N. was understandably slow to respond to his belated appeals for assistance. It also didn't help that Bush's conciliatory tone was widely construed as a manifestation of desperation rather than any genuine change of heart.

Humility Or Humiliation?

Bush's antipathy to the U.N.'s participation in Iraq was politically inane. He initially shunned the U.N. out of irritation at for its reluctance to support our military campaign. Sacrificed in this pique was the wisdom to work with the international organization to defuse the insurgency and spread the huge monetary cost of putting the Mideast country back on its feet. In the cozy recesses of their Washington enclaves, Bush officials agonized more about "saving face" than saving the lives of our soldiers in the field.

The American people deserved better than to be at the mercy of some White House officials' juvenile vindictiveness in slighting the U.N. for its lack of cooperation. When events in Iraq spiraled out of our control, many in the Bush Administration worried that returning to the U.N. hat in hand would be perceived as an act of humiliation, rather than humility.

Whether or not they were right, Bush should have recognized much sooner that making peace with the U.N. would ultimately help heal the breach with many countries that Bush originally alienated during the debate on Iraq.

Cooperation of *all* countries is essential not just to defeat international terrorism, but to protect the earth's climate, preserve its biodiversity, and sustain its marine and other natural resources. Whether Bush likes it or not, the U.N.— flawed as it might be—remains the only vehicle where a collaborative effort of that magnitude can take shape.

Bush Administration detractors of the U.N. may find it hard to stomach, but controlling cross-border pollution requires nations to sign international treaties that demand the sacrifice of some degree of sovereignty for the common good. Bilateral arrangements just won't cut it.

Events in Iraq have compelled Bush to backtrack. But he continues to flirt with transforming us into a rogue state to accommodate corporate chieftains resistant to international regulations, and to oblige ideological unilateralists driven by messianic ambitions of world dominance. It is still possible he will take us there.

Chapter Six

Stealth And Snow Jobs

Loading Up For Bear

President George W. Bush's all-out assault on the nation's environmental protection infrastructure has not been a frontal attack that attracts widespread publicity and quickly galvanizes public opposition. Instead, the president has sought to achieve his objectives behind a smokescreen of rhetoric emphasizing marketplace incentives' superiority to government regulation in improving quality of life. This deference to entrepreneurial activity has served as a euphemism for relaxing, and in some cases eliminating, environmental regulatory restraints on his corporate cronies. These rollbacks have often been achieved through highly technical regulatory changes that go unnoticed by all but the relatively few familiar with the minutiae of the statute.

In addition, environmentalists have seethed at Bush strategists' last-minute, behind-the-scenes efforts to amend major appropriation bills in hopes of winning passage of controversial initiatives without public hearings or a direct congressional vote on the issues.

The administration has also used its broad discretionary authority to chip away at the current regulatory regime. It has delayed compliance deadlines, supposedly for the sake of allowing commercial interests greater flexibility in meeting statutory requirements. Unfortunately, flexibility has all too often dissolved into permissiveness.

The president has attempted to reduce public participation in government decision-making whenever possible, a strategy inherited from his father. George W. may project a folksy, populist charm, but he has the elitist mentality of a board chairman who doesn't take kindly to having his authority questioned.

> *To further obscure from the American people his intent to weaken existing environmental safeguards, Bush has resorted to listing proposed technical but significant changes in the law in the Federal Register late Friday evening, after most news outlets have closed up shop.*

To further obscure from the American people his intent to weaken existing environmental safeguards, Bush has resorted to listing proposed technical but significant changes in the law in the Federal Register late Friday evening, after most news outlets have closed up shop. If you couple the abstruse language of the changes with the late hour of their introduction, you are left with a story line that invites journalistic neglect.

On the advice of GOP political consultant Frank Luntz, Bush has used semantics to weave a web of deception that blurs an unpopular environmental agenda. "Climate change" is substituted for the more menacing "global warming" and the more neutral "conservationist" for the activist label, "environmentalist." [8]

Luntz has other words of wisdom for the president. "In making regulatory decisions involving the environment, the federal government should use best estimates and realistic assumptions rather than worst-case scenarios advanced by environmental extremists." [9]

Bush has taken this counsel to heart, much to the detriment of the elderly, the disabled, the very young, and other fragile elements of our society who need "worst-case scenario" regulations to assure adequate protection from polluters.

Another Bush technique is to make a token effort in court to defend environmental statutes that the president privately opposes. Administration lawyers have become masterful at orchestrating a de facto giveaway to corporate litigants petitioning to weaken environmental safeguards. Where litigation is not involved, the administration has occasionally allowed environmental regulations that it dislikes to either elapse or wither from purposeful neglect.

This statutory sabotage has been enthusiastically executed by Bush appointees, many of whom were previously employed by the very industries vigorously lobbying for environmental safeguards' dilution. In that vein, the president has also used judicial appointments as a backdoor way of subverting environmental protection. He has nominated to the bench a number of individuals with lengthy histories of treating environmental statutes as undue burdens on corporate activity and private property rights.

Justice Triumphs

When Bush junior's Administration has gone to court intent on presenting a winning argument that will weaken environmental protection, it has lost case after case—and the reason is obvious. Our president and his advisors have been trying to make an end run around existing federal environmental statutes. Even most of the judges ideologically aligned with the president have not in good conscience been able to countenance such flagrant shenanigans.

Environmental activist organizations have hauled the administration into court where analytical scrutiny for the first time is brought to bear on the White House's circuitous attempts to undermine the law. It doesn't take long for Bush's subtle assaults to be exposed and crumble under the weight of cross-examination.

Most recently, a court blocked Bush's plan to use his administrative powers to provide industry with more compliance flexibility by weakening the anti-pollution standards under the Clean Air Act. The administration experienced judicial rejection of its initiative to bypass the Marine Mammal Protection Act and allow expanded Navy use of sonar that would harm whales and other creatures of the sea. Its edict (at the behest of commercial interests) to expand air—and noise-polluting snowmobile use in Yellowstone National Park was rescinded by a judge who reinstated the phase-out of the machines as ordered by President Clinton.

In the 94 cases in which the White House took a position that would weaken the National Environmental Protection Act, the heart of the statutory defense against ecological degradation, the administration lost 73 of them. The Bush people also came out losers in 68 of the 76 cases in which, for all intents and purposes, they took industry's side and argued against the Endangered Species Act they were charged with enforcing.

While the American people can be thankful for this judicial oversight, the courts cannot always be relied upon to put the proverbial finger in the dike. There are times when a renegade judge has sided with Bush's insidious attacks on environmental regulation, or when a sound judicial decision has been rendered too late to avert significant ecological damage.

Overthrust Malarkey

President George W. Bush's assertions that the Clinton Administration closed off federal lands to energy development, and that domestic fossil fuel production declined as a consequence, were pure poppycock. In perpetuating this fiction, both in his campaign and after the election, Bush acted as point man for an industry with which he has had longstanding business, political, and social ties.

The truth is that roughly 95 percent of the federal lands that Bush and energy company executives maintained were off-limits to development are in fact open to commercial activity. There was thus no basis for their charges of denied access to vast swaths of publicly owned tracts in the potentially fossil fuel-rich Rocky Mountain Overthrust Belt states of Colorado, Montana, New Mexico, Utah, and Wyoming. Of the 116 million acres of public land in those five states, only 3.9 million were closed to energy extraction, and those areas were primarily federally designated wilderness and national parks.

Nor was Bill Clinton close-minded about energy development, as Bush would have us believe. According to U.S. Bureau of Land Management statistics, the Clinton Administration conducted an aggressive onshore oil and leasing program during its eight-year tenure. Even more tellingly, the total number of bids it actually received was for the most part comparable with those of the previous administrations of Ronald Reagan and George H.W. Bush.

It is fair to say that some of the federal Overthrust Belt lands accessible to industry have restrictions in place to safeguard the environment. The constraints range from seasonal prohibitions to protect wildlife reproduction to limitations on offsite drilling in delicate ecological areas.

One would have thought that an industry which takes pains to insist it is operating in an environmentally sound manner would have welcomed these modest restrictions and used compliance as graphic evidence of its good faith. But there was no need to erect a façade with a sympathetic President Bush in power.

Specious Semantics

Often, George W. Bush's deception has taken the form of cheery environmental platitudes. His administration has paid lip service to strong environmental protections and then resorted to semantic sleights of hand that relieved it of these commitments. Boilerplate environmental declarations have been undercut by cleverly worded caveats, euphemistic code words, subtle omissions, and even subliminal messages warning that one could expect a lot less than meets the eye.

As a Bush surrogate, Interior Secretary Gale Norton has been especially skilled at using semantics to obscure the administration's intention to grease the skids for extractive natural resource industries. She proved that at her confirmation hearings, when she cleverly evaded any firm promises to aggressively enforce existing environmental safeguards, even as she assured senators she would faithfully administer the law.

When asked by senators whether she would oppose widespread drilling on previously protected public lands, she responded that "it is important to have a balance [between environmental protection and economic development]." Guess what "balance" really means! Any senator who accepted a literal translation was hoodwinked.

Queried as to whether she would enforce the federal environmental laws that she had extensively criticized throughout her career, Norton replied that she would *in principle*. It was an answer that left the door ajar to interpreting the laws in a highly arbitrary manner.

Bush himself has been no slouch at verbal obfuscation. While touting the importance of energy conservation, he has been quick to stress that our problems cannot be solved by conservation alone. But none of Bush's critics were suggesting that conservation was the sole salvation for our energy woes. The president was simply creating a straw man to win public support for massive drilling of public lands previously off limits to development.

He also sought to cloud the phase-out of popular (but in his view ideologically incorrect) government programs as he slashed their budgets. To avert criticism, Bush attempted to convince the man in the street that private donations would replace loss of public funding. It is a scenario that was—and is—a hard sell, even to the most naïve.

Arctic Candor

One of Bush junior's most egregious semantic prevarications was to claim that the Arctic National Wildlife Refuge could be developed with minimal environmental impact.

Even if you accepted the president's assurances that new technology would make oil exploration environmentally benign, his failure to mention the effect of oil *production* still undercut his overall contention. The scale of activity in the production phase would dwarf the level in the exploration stage and drastically alter the character of the heart of the refuge. All you ever heard from the Bush Administration, however, was that drilling in the unique ANWR wilderness would be confined to the bleak winter months, when much of the region's wildlife was dormant or had migrated southward. Bush also repeatedly boasted that new diagonal exploratory drilling techniques would reduce the size of energy development to a mere 2000 acres on ANWR's expansive 1.5-million acre coastal plain.

Even his exploration justifications were off-base. Important segments of the wildlife population remain on the Arctic coastal plain throughout the harsh winter, with musk oxen and hibernating polar bears being among the most notable. Exploratory rigs do occupy less space than their predecessors, but Bush was in error when he said that the ice roads required by industry would disappear without a trace on the ANWR tundra upon warmer weather's arrival. Surface water is scarce on the coastal plain, and scar-producing excavation for underground deposits would have to be undertaken to establish the requisite network of ice roads.

As for oil production, it would need to be a year-round operation to have any chance of realizing profitability, and would require an extensive industrial network spreading across hundreds of miles of virgin tundra.

It also turns out that the new fossil fuel extraction technology is not as much of an improvement as the president suggests, at least according to workers in the Prudhoe Bay oil fields on Alaska's North Slope. Their lack of confidence appears to be borne out by the hundreds of oil spills that have occurred every year at Prudhoe, despite the technology's presence.

> *Maybe President Bush genuinely believes that the nation's need to extract whatever oil is interred in ANWR outweighs any risks that development poses to the coastal plain's ecosystem. But if that is the case, why doesn't he have the guts to come out and say so?*

Maybe President Bush genuinely believes that the nation's need to extract whatever oil is interred in ANWR outweighs any risks that development poses to the coastal plain's ecosystem. But if that is the case, why doesn't he have the guts to come out and say so?

Anatomy Of A Scam

George W. Bush's duplicity was in full throttle as he sought to sell his national energy policy to the electorate during the spring of 2001.

First, the president invented a phony short-term energy supply crisis to justify a blueprint heavily weighted in favor of production over conservation. He then tried to use California's energy shortfall as a basis for validating his proposal, but it soon became apparent that the Golden State's quandary stemmed from problems of distribution rather than supply. There was no evidence of an energy crisis across the nation, with fuel price increases and consumption rates substantially trailing non-fuel price increases and economic growth. Bush's crash crusade floundered.

Vice President Dick Cheney also tried to drum up public support by contending the nation faced disaster without construction of 1300 new power plants, an average of one a week over a 20-year period. Yet new power plants were already being built at a clip of roughly three a week, and nearly half of the total Cheney wanted would be on line by 2005. Even more importantly, that number of plants had the potential to fulfill the nation's requirements. The federal government's own energy experts calculated that a combination of energy efficiency initiatives and renewable energy projects would negate the need for building the remainder of the facilities that Cheney was proposing. His vision was clearly overkill, to satisfy the grand designs of the electric utility industry.

Any gasoline derived from Bush's vaunted plan to tap the Arctic National Wildlife Refuge would take 10 years to reach the market. It would also cost approximately 16 percent more than any other domestic source, because of the high transportation costs for delivering the product from such a remote area.

Bush extolled the virtues of fuel efficiency and renewables in his energy plan, but his budgets consistently shortchanged these alternative energy sources compared to fossil fuel and nuclear power.

Trapped

Bush soon found himself caught between a rock and a hard place when it came to environmental protection. If he made a serious attempt to challenge pro-business, anti-big government factions in the name of environmental protection, he would alienate his hard core supporters and betray the philosophy that molded his political career. That left him with no choice but to open himself up to environmentalists' charges that he was in Corporate America's hip pocket.

Religion is a vital cog in George W's life yet, ironically, his environmental policies placed him on a collision course with the convictions of a majority of the world's religious leaders. The president was arguably running afoul of the Bible, which contains such passages as "When you reap the harvest of your land, you shall not reap all the way to the edges of your field" [Leviticus 19:9].

Bad Storytelling

George W. Bush has limited himself to vague generalities in delineating his environmental vision, because he can't afford to be forthright. The president and his aides realized long ago that a strategy noticeably tilting our regulatory system in favor of business interests would not sit well with the electorate. Hence, you hear soothing, scripted bromides designed to obscure the president's allegiance to a narrow ideological agenda.

Operation Stealth

In monopolizing the nation's attention, the war on terrorism provided cover for pro-industry Bush appointees to chip away virtually unnoticed at popular environmental protections. These appointees sometimes pursued their goals through sins of omission, rather than commission, in order to better sneak their initiatives past a distracted citizenry.

Some early examples:

- Weakened environmental safeguards for national monuments newly designated by President Clinton. Bush didn't dare attempt to directly rescind the national monuments designated by his predecessor and wildly popular with the public. Instead, he chose—without any public notice—to modify the small type in the monuments' management rules so as to benefit grazing, mining, and other commercial interests' activities in the supposedly protected areas.
- Deferred implementation of a National Park Service decision to phase out snowmobiles in Yellowstone and

Grand Teton national parks within two years, at the request of commercial interests.
- Interior Secretary Gale Norton did not appeal an Idaho Supreme Court decision denying the Deer Flat National Wildlife Refuge rights to Snake River water. Industry upstream was the beneficiary of this default.
- President Clinton conferred wilderness protection on more than 58 million acres of roadless national forest land, and the initiative received broad public support. But the Bush Administration did virtually nothing to defend the "roadless" policy against lawsuits filed by the timber industry and several states.

Sins Of Omission

The George W. Bush Administration maintains a guarded stance toward full disclosure of any of its activities.

Bush's discomfort with transparency is exemplified by his suppressing, ignoring, or delaying release of facts that detract from industry-oriented objectives. For instance, Interior Secretary Gale Norton was caught red-handed omitting a government study that contradicted her assertion that caribou were not adversely impacted by oil drilling on Alaska's North Slope. When confronted with her omission, Norton smiled sheepishly, shrugged her shoulders, and declared it was simply an honest oversight.

Vice President Dick Cheney is as much a master of withholding information as his boss. We are still trying to find out what role industry operatives, including Enron chief Ken Lay, played in helping Cheney draft the administration's proposed national energy policy.

What was the president's and Cheney's justification for invoking executive privilege to withhold release of conversations and contacts leading to the formulation of their energy plan? Surely the American people were entitled to know the details of government decision-making that would have such a profound effect on their daily lives.

One can't help wondering why the Bush Administration so stubbornly resisted environmental groups' official request to have all the information made public. It was not as if the administration's discussions with energy industry executives dealt with any classified material or military secrets whose disclosure would jeopardize national security. The conversations concerned how best to meet the nation's future energy needs, and what roles conservation, increased fossil fuel production, and environmental regulations should play in the process.

A White House spokesman explained that descriptions of the meetings were withheld from the public because they contained accounts of "discussions inside the administration about what the energy plan should say."

For the life of me, I can't figure out what was so confidential about the pros and cons of oil drilling on public lands, a substantial increase in conservation measures, and other energy policy-related options. Did the Bush people fear that we wouldn't understand what they were talking about? Or were they worried that we would understand all too well? Was the stonewalling an attempt to spare the president and corporate chieftains the embarrassment of being caught collaborating in drafting highly controversial energy policies that subordinate environmental considerations to drilling and mining?

If industry lobbyists truly believed a relaxation of environmental regulations were necessary to assure future ample energy supplies, what was wrong with openly making their case to the American people?

President Bush and his energy task force leader, Vice President Dick Cheney, should not conceal the give and take of policy-making, whether through executive privilege, technical exemptions to the Freedom of Information Act, or any other strategy. Full disclosure of the deliberative process is always the preferred alternative in a democracy, with any suppression of open debate (except when national security might be compromised) nudging our political system towards authoritarianism.

Moreover, knowing the identity of who was making a particular argument could be very important in weighing the credibility of that argument. For example, the American people ought to know beforehand if a deregulatory provision in the Bush plan was the brainchild of a corporate magnate who would reap huge profits from a rollback of environmental rules.

Vice President Cheney has insisted that his energy task force's deliberations remain confidential to protect the president's need to receive "candid advice."

Why aren't the American people entitled to hear this same "candid advice" when it could have monumental bearing on their physical health as well as their pocketbooks?

Litigation is underway and, hopefully, the U.S. Supreme Court will compel Cheney to abandon the fig leaf of executive privilege and come clean.

In The Eyes Of The Beholder

Saddam Hussein was not the only one manipulating scientists to promote an agenda. George W.'s Administration rigged federal government scientific advisory panels with individuals favoring the president's views and prepared to act as a rubber stamp.

To pave the way for such an arrangement, scientists sometimes had to be subtracted as well as added. A case in point was a highly qualified University of Massachusetts researcher, whom the president rejected for an occupational health panel. The unspoken reason: she had publicly supported a workplace safety regulation that Bush repealed the previous year.

The president has also packed supposedly politically impartial panels with conservative ideologues ready to exploit the degree of inconclusiveness that exists in virtually every scientific finding (especially an unfavorable one). Science is a constant work in progress, and thus rarely can attach any finality to its conclusions. That creates an opening for Bush's scientific henchmen to inject the issue of uncertainty and derail any tough new environmental regulations opposed by the deregulatory-minded administration.

The president does not want scientific advice; he wants affirmation of pre-determined conclusions, and that's not what the panels are supposed to do. Their designated mission is to provide information and recommendations for formulating the most soundly crafted environmental regulations possible.

Still, Bush's tactics were no surprise. They were consistent with his persistent attempts to redefine "best available science" in terms of furthering his agenda.

"Best available science" is traditionally regarded as a scientific consensus based on the weight of existing evidence resulting from extensive peer review. But the Bush Administration has frequently chosen to ignore overwhelming scientific consensus that does not square with a desired result. Instead, the conclusions of a few dissenting scientists who share the president's view are treated as gospel.

Dollar Deception

President Bush's 2004 Energy Department budget request was advertised as setting us squarely on the path towards a clean, renewable energy future. Yet the bulk of the budgetary increase for renewables went to development of hydrogen-powered fuel-cell cars, a noble objective but one at least a decade away. Available renewable energy technologies—solar and wind—were given short shrift. So were energy efficiency programs. White House officials offered the explanation that the private sector had harnessed solar and wind technology and needed little in the way of government help. The president obviously felt differently about the well-established nuclear and coal industries, which benefited handsomely from Washington's largesse.

The Patriotism Factor

When advocacy of policies lacking the support of the majority of Americans generated too much unfavorable publicity in the aftermath of 9/11, the administration was not shy about playing the patriotism card. Hence, national security was invoked as reason to cover (and despoil) the Arctic National Wildlife Refuge's ecologically singular coastal plain with oil industry infrastructure.

Chafing at restrictions imposed by the Endangered Species Act, the administration alleged that the popular environmental law impeded military training exercises on some of our homeland bases and demanded exemptions from the statute. No matter that it was well documented that environmental protection did not adversely impact either national security or troop preparedness at those sites. At least President Bush stopped short of equating environmental activism with treason.

Trust Us

Empirical evidence suggests two major rationales behind President George W. Bush's penchant for secrecy. Having grown up in a corporate environment where chief executives of companies were often laws unto themselves, Bush appears to have entered the Oval Office with a paternalistic attitude toward the public.

Despite his folksy Texas twang, Bush is more of a patrician than a populist. You can almost hear the underlying message emanating from the White House: "Don't trouble yourself with numbers or policy formulations that you probably won't understand and could change at a moment's notice anyway. Just let the president do his job, and you won't be sorry. Trust us."

There is another reason for Bush's clandestine modus operandi. Silence is golden when one has something to hide, which is why the president and his team would rather not draw attention to highly ideological policies they know would have difficulty garnering majority support. The hope is that if the policies can be slipped past Congress and go into effect, the public will realize the error of its ways and appreciate the president's foresight. Thus, we have the administration withholding information on everything from the cost of our involvement in post-war Iraq to who participated in the drafting of Bush's proposed controversial national energy policy.

Summer Doldrums

There was something unseemly about George W. Bush (or any president, for that matter) hosting one fund-raising barbecue after another while the troops he sent to Iraq were suffering combat casualties on a daily basis.

But that was not the only disquieting display by Bush during the 2003 summer hiatus from Washington. He traveled to several bucolic locations where he attempted to embellish his shaky environmental credentials with flowery speeches and staged photographs.

Bush blissfully gushed to reporters about restoring native grasses and observing the arrival of a family of bobwhites on his Crawford, Texas, ranch. He is to be commended for his communion with nature. But in the meantime, an estimated 34 animal and plant species outside the friendly confines of his ranch went extinct on his watch. Moreover, his administration listed 24 species as endangered when, during an equivalent period of time, the Clinton Administration designated 211 species and even Bush's own father classified 80 for federal protection.

The president reveled in the biological diversity on his Texas spread, yet his administration was being sued for allowing continued use of a highly toxic herbicide (Atrazine) that jeopardized the survival of endangered species and other wildlife on public as well as private lands.

Speaking before cameras during a ceremonial August 15 visit to the Santa Monica (California) Mountain National Recreation Area, Bush declared that "an important point about conservation is man's ability to make sure God's beauty is nurtured and preserved."

He sure had an odd way of practicing what he preached. His top priorities included opening up the ecologically priceless Arctic National Wildlife Refuge to environmentally degrading oil drilling and removing wilderness protection from 58 million acres in our national forests. When he delivered the Santa Monica speech, he had already cancelled temporary protection of nearly a quarter-million acres of wilderness study areas in the Western United States. "God's beauty" stood to take a beating.

Transparency

When President Bush recently declared with a straight face that he believed in "transparency," I nearly choked on my Cheerios. Bush was being interviewed on an Arabic news program regarding our military's abuse of Iraqi prisoners. In the course of the questioning, he said, "We have nothing to hide. We believe in transparency. We are a free society."

> But, as we have seen when it comes to Bush's handling of the environment, much of his activity has been cloaked in secrecy.

To what degree that transparency will come into play in the investigation of the treatment of Iraqi detainees has yet to be determined. But, as we have seen when it comes to Bush's handling of the environment, much of his activity has been cloaked in secrecy. Transparency, to use the military vernacular, has been "missing in action." Hence, Bush's claim that his administration is an open book looks to be either a product of self-delusion or a deliberate distortion of reality.

Chapter Seven

Blunders And Blind Spots—
A Prisoner Of Ideology

The American Way

President George W. Bush argues that his "drill first and ask questions later" energy policy is essential to protect the "American way of life." Just what does he mean by the "American way of life"?

All indications are that the president is referring to the "wide open spaces" mentality prevalent among our pioneer ancestors as the nation expanded westward. America was viewed as an endless cornucopia of natural resources to be exploited, with no thought—or need to worry—about sustainability. There was nothing pejorative about engaging in ostentatious behavior. Indeed, the more one could afford to indulge in excess, even waste, the more one was considered successful.

New York Times columnist Frank Rich aptly described the modern practitioners of this lifestyle as those who "want what they want when they want it and are utterly convinced of their righteousness in thinking that way."

It's not surprising to hear people with that sort of attitude maintain that the ability to pay for environmentally detrimental goods or services validates the legitimacy of such transactions.

The combination of this gluttonous predilection and our good fortune in having a large, resource-rich country has resulted in the following: Although we constitute only five percent of the global population, we gobble up approximately 30 percent of the world's energy and are the source of 24 percent of human-generated polluting carbon emissions. If these statistics prick your conscience, you may become even more restive at the disclosure that, as individuals, we use almost twice as much energy per person than our European and Japanese counterparts, despite their enjoying nearly identical standards of living.

Our affinity for overindulgence and our apathy at the consequences have produced behavior that has led to the loss of more than one-half of the lower 48 states' biologically rich wetlands. It has resulted in the surface area covered by live coral on the largest reef in the Florida Keys shrinking from 50 percent in the 1970s to only five percent today. The average annual distance Americans drive has quadrupled over the last 50 years, even though development in many cases has brought our destinations closer to home. Air quality has suffered accordingly.

With modern civilization's advances has come the realization that our planet's life support system does have limits. Soon, it will be impossible for even the most wedded to the consumptive status quo to ignore the question of whether a bulging billfold gives one carte blanche to do as one pleases.

Those unwilling to abandon our unrestrained, consumption-oriented lifestyle have come up with a rationalization to support their stance. They contend that technological innovation will remedy any abuse we inflict upon the natural world. But technology has given us no reason to believe it can restore complex, often delicately balanced ecosystems on which all life on earth depends.

We may reject critics in other countries telling us how to exercise our freedoms, and perfunctorily dismiss as sour grapes their complaints that we have been too heavy-handed with the earth in amassing our wealth. But what about our obligation to succeeding generations of our own flesh and blood? Should we risk bequeathing them "wide open spaces" drastically diluted of life?

A Shaky Pillar

Sound science is a basic pillar of President George W. Bush's environmental policy, and it is on shaky ground.

Critics have duly noted Bush's hypocrisy regarding "sound science" in connection with the global warming and missile defense issues. He has urged full speed ahead with construction of a missile defense shield, even though scientists have yet to determine whether it can work. Meanwhile, the president is advocating a "go slow" approach in combating global warming, despite years of research producing a broad scientific consensus that the problem is real and demands immediate attention.

Justice Be Served

President Bush says that, when he gets the chance, he wants to appoint to the U.S. Supreme Court a conservative justice in the tradition of Antonin Scalia and Clarence Thomas. They are jurists, Bush adds, who interpret the law, not make it.

Obviously, the president hasn't gotten the word that his two favorite Supreme Court justices are far from models of judicial restraint, at least in regard to certain environmental laws.

Some examples of Scalia's and Thomas' handiwork:

- They have broken long-established precedent by interpreting the Constitution's Commerce Clause in a manner that strips the federal government of the authority to regulate environmental abuse and instead transfers the power to the states, some of which are not up to the task. It is a reflection of the judges' hostility toward "big government" and their predisposition toward "states' rights."
- They have construed the Fifth Amendment to require

the public to pay corporations for complying with environmental laws, on the grounds the companies would otherwise be deprived of the use of their property without compensation. It is in essence paying polluters not to pollute, certainly a boon to industry but not a directive supported in the actual wording or legislative history of environmental law. These are rulings of judges who believe that private property rights should take precedence over the public trust, rather than be balanced against it. Forgotten is the elementary principle that any governmental regulations imposed on a private property owner to prevent harm to other property owners do not warrant compensation.
- Justices Scalia and Thomas have sought to narrow grassroots environmental activists' standing to take alleged polluters to court, whereas the two jurists have vigorously defended corporate polluters' right to sue the federal government for alleged excessive regulation.

What about "liberal" activist judges? Don't they also take liberties in interpreting the law? In regard to environmental statutes, they really don't need to. The language in the Clean Air Act, the Clean Water Act, and other environmental laws clearly spells out that environmental protection must receive top priority.

Beyond Ideology

> As misguided as most of his ideologically driven environmental policies are, President George W. Bush will only change course when events leave him no choice.

As misguided as most of his ideologically driven environmental policies are, President George W. Bush will only change course when events leave him no choice.

His stubborn refusal to look beyond a narrow set of preconceived notions until compelled by circumstances is not a prescription for responsible leadership. What we are seeing is a reaction to environmental problems after the fact, rather than a proactive strategy to prevent or mitigate them before they ratchet into high gear.

It took an act of toxic terror to bring the president back to discussions from which he had withdrawn two months earlier, regarding the renegotiation of the 1972 Biological Warfare Treaty. The anthrax contamination of some of our mail persuaded the president to discard his unilateral protectionist stance and rejoin the community of nations in trying to hammer out an effective enforcement mechanism for the global ban.

Timidity From A Tough Guy

President George W. Bush's plan to combat global warming is fine as far as it goes. Unfortunately, it does not go nearly far enough.

His proposal is rooted in the principle that "economic growth is key to environmental progress, because it is growth that provides the resources for investment in clean technologies."

There is some truth to that concept, provided the economic growth is environmentally sustainable in character. Clean technologies alone cannot save an environment that is deliberately and rampantly polluted.

Bush never spells out the sort of economic growth he envisions—and with good reason. His administration is known to be contemplating moves to accelerate economic growth by reducing pollution abatement requirements for power plants and easing environmental restrictions on industrial activity in publicly owned conservation areas. It's an idea that is a surefire formula for realizing short-term profits at the expense of long-term public—and environmental—health.

It's The Science, Tupid!

For someone who has vowed to base his policies on the best science available, President George W. Bush is giving surprisingly short shrift to research in a number of important fields.

In the biotechnology world of stem cell research, Bush's decision to limit federal funding only to experimentation with existing cell lines is creating a situation in which the United States could easily fall behind Europe in medical advances. Indeed, Europe is increasingly becoming the place for scientists to engage in original stem cell research. They are being lured by a more lenient stance toward the creation of *new* stem cell lines. While the morality of Bush's stance is being debated in our nation, the enormous potential of new cell lines to cure some of the most intractable diseases is being energetically pursued as a matter of public policy across the Atlantic.

Due to the Bush Administration's lukewarm support of research and development of clean, renewable energy resources, we are lagging in that sector as well. In his latest budget proposal, the president has slashed funding for all renewables by one third, and for solar energy specifically, by nearly one half. It is hardly an opportune time for this miserly approach, considering that the latest statistics show Japan produced 171.22 megawatts of solar photovoltaic products in 2001, compared to our 100.32 megawatts. And to think, we were once the pioneers in this technology! By the way, fossil fuel research endures no such fiscal hardship.

The president is asking for a 15 percent reduction in the U.S. Forest Service's fish and wildlife research budget. It is an odd request for a political leader seeking to make a strong case for expanding commercial exploitation of our national forests without harming the environment.

These actions are not representative of the way Bush promised to utilize research. His cherry-picking for the outcome he desires has some scientists muttering, "With a friend like the president, who needs enemies?"

Cockeyed Priorities

While President Bush has been quick to sign off on a host of arguably duplicative multi-billion-dollar weapons systems, he is quite ready to cut corners with bee research, by leaving only one center operative (in Texas, of course). Even then, he would slash the facility's $5.7 million budget by more than half and reduce the number of positions from 21 to nine.

Among the facilities that the president would close is the Baton Rouge lab, where work is being done to breed strains of honeybees resistant to parasitic mites. Also on the hit list is the Beltsville, Maryland, bee research center. Scientists there are seeking to develop biological pest controls that would replace the highly toxic pesticides many beekeepers are reluctant to use for fear the cure would be worse than the disease.

It clearly has not hit home to Bush that production of approximately one-third of our food is dependent on pollination, primarily from honeybees. It's the intensity of pollination that determines the yield and quality of such fruits as apples, pears, cherries, and plums. California, which provides much of the nation's produce, is almost totally reliant on honeybees for fertilization of its crops.

The Bush Administration's misplaced priorities shouldn't come as a surprise. This is the same bunch who don't see anything wrong with shearing off a mountain top to get at a seam of coal, and then permanently contaminating the streams and rivers in the valleys below by dumping mining waste into the waterways. For a one-time use of the coal, the White House is willing to sacrifice an endless supply of fresh, clean water.

The president says we can have both a healthy environment and a robust economy, but he often acts as though they were mutually exclusive.

His inconsistency stems from paying lip service to the politically correct notion that it's possible for the two objectives to co-exist in our modern society, but failing to acknowledge—and act upon—the more urgent reality that, without each other, the two ideal states ultimately *cannot* exist in sustainable fashion.

The False Populist

President George W. Bush seeks to pass himself off as a populist, but his actions have been those of a plutocrat. Virtually at every juncture, Bush's environmental policy has favored corporate interests over the public interest. You might argue, as Bush and his followers do, that corporate and public interests are interchangeable. And it is true enough that industry leaders are part of the general populace. But the idea that corporate good fortune is automatically shared with the public at large—the so-called "trickle down" theory—is a false premise. "Trickle down" has the capacity to raise false hopes and is often used to justify corporate America receiving preferential treatment that leads to environmental degradation.

President Bush's policies have aimed at making it easier to open up previously protected, undeveloped public lands to commercial exploitation. He has sought to roll back the pollution abatement requirements for heavy industry so that business leaders can lower overhead costs and maximize profits.

Let's be candid about what's going on in the White House these days vis-à-vis environmental policy. It's not the man in the street but the man in the executive suite who is getting most of the attention.

Volunteerism Copout?

One hates to be suspicious of an outwardly admirable Bush Administration initiative to create a government program in which volunteers will be recruited to clean up and maintain public lands. Certainly, it's a great idea if the volunteers are a supplement to, rather than a replacement for, the dedicated corps of career public servants engaged in managing our country's natural resources.

Unfortunately, there is concern that the administration's recently unveiled "Take Pride in America" volunteer program is meant to provide substitutes for the federal workforce assigned to conservation tasks. It's a move that would be consistent with the budgetary squeeze brought on by the Iraq war and George W's supposed ideological aversion to "big government" and sprawling federal bureaucracies. There is no question that the White House is looking to cut domestic budgetary corners to compensate for the escalating costs of operating our enormous military machine. To save money and mitigate the exploding national budget deficit, the administration can be expected to resort to volunteerism with ever-increasing frequency.

But what about the professionalism, reliability, dedication, and long-term commitment for which park rangers and other federal employees in the conservation field are famous? No knock on volunteerism, which is a noble endeavor, but ultimately you get what you pay for.

As an added inducement to attract volunteers for the great outdoors, Interior Secretary Gale Norton pledges that "outstanding efforts will be rewarded with presidential recognition." That sounds to me like the presentation of some sort of plaque, which is a heck of lot cheaper than doling out an annual salary to a competent federal employee with years of expertise.

Health Versus Profit

As Bush's term begins winding down, evidence is mounting that he is more concerned about protecting corporate profits than public health.

In one of his administration's most unconscionable acts, we have joined with the Dominican Republic to become the only two countries out of 173 to balk at signing a treaty to curb the sale and use of tobacco products throughout the world. (Just before this book went to press, we did sign, but our ratification of the pact is by no means assured.) Under the accord, tobacco consumption would be discouraged by more conspicuous labeling, higher taxes, and bans on advertising when not in conflict with national laws. Our government paid lip service to the treaty's benevolent objective of reducing the estimated 4.9 million yearly tobacco-related deaths around the globe. But in our next nicotine-free breath, we insisted on inclusion of a provision that would, in effect, give signatory nations the discretion to exempt themselves from any or all of the treaty's requirements.

Want to guess why we were making this demand? Let me help. Increasingly losing face (and business) at home, American tobacco companies have begun to concentrate their merchandising efforts on foreign markets, where the public is not nearly as well attuned to cigarettes' hazardous health effects. President Bush feels the tobacco kings' pain, and he is doing something about it.

> *In another bow to business interests, the Bush Administration has issued workplace safety rules that rely heavily on industry's voluntary compliance.*

In another bow to business interests, the Bush Administration has issued workplace safety rules that rely heavily on industry's voluntary compliance. It is a reactive rather than pro-active regulatory approach that only kicks in after damage to workers' health has already occurred. The idea is to spare industry any extra compliance costs except in an emergency. A company taking remedial action after the fact loses some bucks (which it usually has a good chance of recovering), while the victimized workers lose their health and possibly even their lives.

Finally, the administration has been reluctant to set anti-pollution standards beyond levels adjusted to protect healthy young adults from potentially harmful exposures. Bush's budgetary planners have agonized over providing an extra degree of protection to the most vulnerable in our society—the elderly, the infirm, and the very young. Why the hesitation? The extra protection would save relatively few lives, at a far greater expense to industry. Besides, the old and the frail don't have all that long to live, and the very young are a lot more resilient than we think.

Denial

George W. Bush's Administration's refuses to acknowledge global warming as a major environmental problem requiring immediate remedial action. This stance represents a dangerous gamble with the future of the American people—and the world.

By choosing to downplay a potentially adverse change in climate resulting from human activity, and delaying the start of any corrective measures, the White House is violating a sacred trust it won at the ballot box. And the course it has adopted is doubly reprehensible, given that we are the leading emitter of greenhouse gases linked to the global warming phenomenon.

The administration's decision to omit any reference to a global warming threat in its *Draft Report on the Environment* reflects its ideologically motivated denial of the scientific consensus that humanity faces a potentially disastrous ecological problem. Instead, White House officials have flirted with the self-serving contention of some ultra-conservative business interests that global warming is really a left-wing environmentalist plot to justify regulation that undermines free enterprise and fossil fuel industry expansion. This school of thought argues that human-induced global warming is no more than an unproven theory and, at most, merits further study. It's a convenient posture to take for those who are dead set against spending more money on pollution abatement and want to postpone such an obligation for as long as possible.

Bush dutifully accepts the premise that we need to go slow because science has failed to conclusively prove environmentalists' claims that global warming demands prompt corrective action. But science rarely proves anything. It is a tool for building a hypothesis based on the weight of evidence. The hypothesis—along with other considerations, such as ethical and political concerns—is then used to formulate policy. From the weight of evidence, the overwhelming majority of the scientific community has deduced that global warming is a legitimate threat, exacerbated by human-generated fossil fuel emissions, and should be subjected to a pollution abatement strategy without delay.

Now is the time for President Bush to transcend ideology and abandon a policy of procrastination. The American people deserve no less.

Your Kinda Guy?

Many "greens" were alarmed at the prospect of Utah Governor Michael O. Leavitt becoming President Bush's new Environmental Protection Agency administrator.

But there was no need to worry, at least if you believed that, all things being equal, the private sector did a better job than the government in protecting the environment.

You shouldn't have gotten apoplectic about Bush's EPA nominee if you were convinced the states were normally better suited than federal governmental agencies to oversee regulation of the environment (though many of the most serious pollution problems are trans-boundary in scope).

Leavitt's nomination should have been reassuring if you were confident that corporations' cooperation in curbing pollution was gained more effectively through voluntary responses than through regulatory mandates carrying stiff penalties.

There was reason to rejoice if you thought that conservation had been favored over industrial development in regulatory environmental decision-making, and a "balance" had to somehow be restored.

Leavitt was a nominee for the ages if you regarded cost-benefit ratios as the overriding principle, rather than simply one of a number of factors behind the formulation of environmental policy. Dollars and cents computations were the fairest way of determining who lives and who dies, right?

If you thought our nation's future depended on extracting every bit of recoverable oil and gas on public lands, even in environmentally sensitive or pristine areas under consideration for protection by Congress, you were comfortable with Leavitt at the EPA helm.

Suppose there was no question in your mind that a vast tract of natural wetlands was expendable when blocking the proposed route of a major new highway. Take comfort in knowing there is a person with a similar mindset running the EPA show.

You are someone who feels deliberations and negotiations on environmental policy are too delicate to be exposed to public scrutiny and should be held in secret. Not to worry. The Utah governor will do his best to put your worries to rest.

Gagging On A Rule

By dutifully adhering to conservative Christian fundamentalist ideologues' misguided demands to cut off funds to overseas non-governmental health service organizations that in any way countenance abortion, even if only as a last resort, President Bush has produced results directly contrary to his intent.

His zeal to curb abortion abroad in the name of "moral purity" has condemned hundreds of thousands, perhaps millions, of Third World women and children to unnecessary suffering and premature death.

George W's specific instrument for unintentionally spreading affliction is the global gag rule, rescinded by President Clinton but reinstated at the behest of the GOP's hard-core conservative base. The gag rule prohibits American family planning financial assistance to any foreign non-governmental health organizations that include abortion as a possible option in treating reproductive health problems. It makes no difference if abortion is *legal* in the developing nation in question.

What we are seeing is the same mentality that drove Bush to require one third of our overseas AIDS prevention funds to be used to promote abstinence before marriage in countries where pre-marital and extra-marital sex have long been accepted practices.

It's all right to promote your beliefs, but to try to *force* one's values overnight on a different culture is a recipe for disaster. Money that could have been used for contraception and medicine to save lives is instead wasted on a message likely to fall on deaf ears. Many people overseas, especially from cultures different than ours, don't make good guinea pigs for conversion when it comes to sexual mores.

Denied the American financial assistance needed to fulfill their humanitarian missions, numerous overseas organizations have been forced to operate without sufficient family planning counseling and contraceptive distribution services. Unwanted and unplanned pregnancies have increased, and this is turn has led to *more*, not *fewer*, abortions.

One can only hope that, sooner rather than later, President Bush comes to his senses and realizes that his strategy to save the lives of the unborn is only reducing the survival chances of the living.

The Web Of Life

The George W. Bush Administration seems more intent on rolling back the Endangered Species Act for the sake of entrepreneurial gain than upholding the law to save rare plants and animals.

As you might expect, any species seems expendable if it ends up in some way interfering with a company's plans for expansion. The president and his minions exhibit no signs of any appreciation for the interconnectedness of nature and do not appear to assign any intrinsic value to God's creatures. For an administration that wears religion on its sleeve, its dismissive attitude toward the wonders of Creation is a glaring inconsistency, to put it mildly. Indeed, if religion is at issue, the White House seems most beholden to Mammon. Its proclivity for weakening the ESA and other environmental laws at the behest of corporate campaign contributors testifies to that.

Under Bush's aegis, only eight species a year have been listed as endangered, and their designation was the result of court orders compelling the president to take such action. Compare his record with Clinton, who initiated an average annual listing of 65 species. The president's own father voluntarily listed 58 or so a year, and even a deregulatory-minded Ronald Reagan averaged 32 species annually.

In President Bush's view, "what's good for General Motors is good for the country." As anyone with environmental sensitivity can attest, that ain't necessarily so.

The Bush Rationale

George W. Bush heads an administration that operates from the premise that "the end justifies the means." It could be no other way. Bush and his advisors know the majority of Americans don't share the president's enthusiasm for environmental and energy policies rooted in corporate cronyism.

Despite these odds, the administration remains confident that if the president's business-oriented approach is given half a chance, the dissenters will realize the errors of their ways. Hence, any deception (short of a prosecutorial offense) that furthers the president's ideological agenda and prospects for reelection is a welcome addition.

One can only note the supreme irony of an ultra-conservative president evoking Marxist doctrine to advance his agenda.

Chapter Eight

Worse Than Reagan?

Divide And Conquer

President George W. Bush has used his would-be signature domestic issue—job creation—to try to undermine the traditional alliance between environmentalists and organized labor.

He maintains that his energy initiative, of which a major component is opening up the Arctic National Wildlife Refuge to oil drilling, would be a prolific generator of jobs. By the same token, he warns that a mandated reduction in greenhouse emissions, such as that contained in the environmentalist-supported Kyoto global warming treaty, would result in a devastating loss of jobs. It is an argument that he and his chief political strategist, Karl Rove, hope will erode a solidarity formed by the two factions over the years out of mutual concern about industrial pollution's adverse effects.

The environmental community and organized labor have responded by giving Bush a dose of his own political medicine. They are offering an alternative energy plan (excluding ANWR) that they contend will produce a *net* gain of 660,000 jobs by 2010 and 1.4 million jobs by 2020. At the same time, they maintain that, within a decade, their energy plan can reduce carbon emissions by 27 percent below projected levels, and by the year 2020, the reductions will amount to 51 percent. It's a clean-up that dwarfs Bush's projections.

How would they pull this off? Their plan relies significantly upon the always politically contentious introduction of a new tax. In their case, it would be a modest carbon tax on energy-producing sources. To make the levy more politically palatable, environmentalists and labor stress that the arrangement is simply making sure that polluters (as opposed to the general public) are fiscally liable for the contamination they generate. And for those who would argue that such a levy would be a drag on the economy, the enviro-labor coalition points out that the revenues from the impost would be returned to the public in the form of cuts in taxes on wages and financing for transitional programs for displaced workers.

Hoodwinking The Clergy

President George W. Bush is a religious man, but there is evidently a limit to his piety.

Religious leaders sought a meeting with his administration in the spring of 2002 to make a scriptural case for stronger environmental activism on the White House's part. They requested an audience with presidential environmental advisors, but they were met by two mid-level employees from the President's Council on Faith-Based and Community Initiatives (FBCI). There wasn't a single environmental official in attendance, not even a token low-level functionary from any of the relevant federal agencies.

The FBCI officials listened sympathetically to the clergy's environmental appeals but were not intimately familiar with the issues. And why should they be? The FBCI was formed to focus on social welfare problems, not environmental ones.

This delegation of clerics received a symbolic cold shoulder because Bush knew what the theologians were going to say, and it was not what he wanted to hear. Appeals for halting environmental assaults on wilderness territory and other natural resources didn't fit in with his game plan of increasing industrial activity on federal lands and relaxing environmental regulations, purportedly to stimulate economic growth.

Campaign Potholes

Much of George W. Bush's tilt toward commercial exploitation of public lands has been occurring in the Western United States, where the shifting demographics do not favor the president's environmental philosophy. An opportunity is emerging for Democratic candidates in a region where extractive natural resource industries are rapidly being replaced by a high-tech and tourist-driven economy. In addition, the West's population has become more concentrated in urban areas than in any other part of the nation, giving conservation an extra leg to stand on when in conflict with proposed commercial activity on scenic federal lands.

Under Suspicion

When questioned about corporate malfeasance, President Bush vows to hold business leaders accountable for their actions and throw any violators of the law into jail. But if his noble pronouncements continue to be accompanied by "giving away the store" to corporate polluters and natural resource extractive industries, Bush's tough rhetoric is not going to help him with the electorate.

Considering Bush's past involvement in a corporate culture replete with questionable financial wheeling and dealing, you would think he would seek to purge even the appearance of corporate favoritism in his policy-making. Yet his conservative environmental philosophy is so unalterably grounded in pro-industry bias that he appears trapped in a political box. He would have to do a complete ideological about-face that would alienate his hard-core supporters and evoke suspicion among the uncommitted.

> ... Bush is headed inexorably toward opening up previously undisturbed public lands to resource extraction companies, especially oil drillers.

Thus, despite corporate America's badly tarnished image, Bush is headed inexorably toward opening up previously undisturbed public lands to resource extraction companies, especially oil drillers. He has moved to give commercial interests more access to publicly owned wildlife habitat, wetlands and watersheds in the name of economic development.

Crocodile tears are rampant in Bush's world. Corporate executives who complain that stringent new anti-pollution regulations will ultimately drive them out of business have been saying that for years, as their companies' stock prices have soared.

This is the bunch with whom President Bush has thrown in his lot!

Poll Watching

The Republican Party made a cottage industry out of criticizing President Clinton for shaping policy solely on the basis of public opinion polls. It's ironic, then, that President George W. Bush may get himself into hot water for not paying enough attention to them.

A 2002 Zogby nationwide poll demonstrated how much the president's permissive "Clear Skies" program was out of sync with public sentiment. Respondents to the questionnaire overwhelmingly favored the government setting specific standards to reduce industrial greenhouse carbon dioxide emissions. They believed immediate steps to curb global warming were essential, and that renewable energy and more fuel-efficient vehicles should be major tools in achieving that objective. There was little faith in the corporate world cooperating voluntarily. All this ran counter to

Bush's actual policy, and you have to wonder when these polls will catch up with the president.

Terrorizing The Environment

Much of the war on terrorism will focus on guarding against our air, water, and crops being sabotaged through the use of biological or chemical agents.

Expanded use of renewable energy is an excellent strategy for reducing terrorists' opportunity to disrupt our electric power delivery systems. Destruction of windmills or solar panels wouldn't release any lethal emissions or produce massive fiery explosions, whereas the sabotage of a nuclear or fossil fuel power plant could. Furthermore, most renewable energy sources are decentralized rather than concentrated in a single power grid, making them far less vulnerable to a crippling terrorist attack.

Other important environmentally related strategies vital to an effective counter-terrorism effort include:

- Keeping the Freedom of Information Act sufficiently unfettered, so that the public can exercise its "right to know" which local industries are lax in their handling of toxic materials and need to be pressured to clean up their act.
- Heightening security at power plants, water utilities, chemical factories, and other sensitive public places.
- Instituting safer technology and more secure disposal of any hazardous wastes at the aforementioned facilities.
- Strengthening public health agencies' response to any terrorist-induced contamination of the environment.
- More careful monitoring of our food supplies, including stricter surveillance of our borders to guard against deliberate introductions of pathogens. We also need to improve our intelligence information-gathering on the

enemy's intentions, and accelerate research to develop disease-resistant plant strains.

Finally, many of the roots of terrorism can be found in the squalor of developing countries. Without clean water, decent sanitation, and productive soil, people often feel they have no future and lash out angrily at anyone who does. Thus, an important step is to upgrade abysmal environmental conditions that make Third World countries fertile ground for recruiting terrorists.

The Bush Administration has become more attentive to security around our strategic public facilities. But it has failed to make the crucial connection between maintaining a healthy natural environment and safeguarding our long-range national security. Instead, it has been using the terrorist threat as a pretext to pursue a rigid ideological pro-business agenda in which there is a bias against federal environmental regulations. The president characterizes many of these regulations as barriers to entrepreneurial activity and hence strategic threats to the nation's economy.

It would be ironic if we, rather than terrorists, turned out to be the worst enemies of our environment.

Supreme Irony

The supreme irony of the George W. Bush Administration is that the president has sought to institute policies that run counter to the philosophical principles on which he ran for office.

Bush campaigned for the presidency on a platform emphasizing the downsizing of big government and the delegation of more of Washington's authority to states and localities. What has happened since those halcyon days on the campaign trail is that Bush has actually moved to concentrate greater power in the White House, frequently invoking the war on terrorism as justification. This wouldn't be so bad if the expansion of the federal role served to strengthen environmental protection, but alas, it does just the opposite. A manifestation of Bush's power grab at the expense of the environment is his attempt to weaken the National Environmental Policy Act. For more than 30 years, NEPA has been a legislative mainstay of environmental protection, requiring federal agencies to review the environmental impacts of their actions and include the public in their decision-making.

The Bush team is too clever to unleash a frontal attack on such a popular law. Instead, the president has sought to undermine NEPA by giving the heads of federal agencies discretionary authority to accelerate and even circumvent the review process if "appropriate" (a.k.a. carte blanche in the name of "national security"). He would reduce the opportunities for public participation in the NEPA deliberations, thereby consolidating even more power in the executive branch.

When states and local jurisdictions don't see eye to eye with him, Bush has invoked the war on terrorism as justification for abandoning his long-declared commitment to give them more say in federal decisions impacting their territory. In the administration's eagerness to open up previously protected public lands to oil and gas drilling, it has on occasion ignored the reservations if not outright opposition of local interests concerned about environmental damage. Once again, national security has served as a convenient excuse to further the president's agenda.

Finally, contrary to Bush's frequently expression of confidence that the workings of a largely unfettered marketplace are best able to arrive at the soundest economical and environmental solutions, his administration has doled out generous federal subsidies (labeled in some quarters as "corporate welfare") to the fossil fuel industry. By contrast, clean renewable energy technologies enjoy no such equivalent government largesse, leaving them unable to compete in the marketplace on a level playing field.

Game Plan

What can environmentalists do to thwart a president who displays ideologically divisive hostility towards most progressive environmental reforms and has regularly sought to weaken anti-pollution regulations?

It is imperative that, when confronted with a proposed Bush initiative that undermines environmental protection, activists should refrain from showering the president with epithets. They should concentrate on the issues, discuss the pros and cons with clarity, and let the facts speak for themselves. The public will come running, no matter how well liked the president might be at the time. If you're skeptical about that, recall how the exposure of Ronald Reagan's destructive environmental policies brought them into widespread public disrepute, despite his personal popularity.

Also remember that it would be difficult to truly savage Bush junior's environmental motives in the public's eyes. Who is going to believe that the president would knowingly risk harming the environment? Besides, Bush and his ideologically motivated advisors genuinely believe that commercial development of public conservation areas and rolling back environmental regulations are in the country's best interests.

Well, a clean environment is good business!

What's their rationale? They are convinced that a strong economy is a prerequisite to a healthy environment. Their reasoning is that the profit motive is the most effective tool for cleaning up the environment because a clean environment is "good business." Well, a clean environment *is* good business! The only problem is that the full economic benefits of pollution abatement or resource protection often accrue over time. Many, if not most, businesses are judged by their success in realizing immediate profit so, if left to their own druthers, the profit they might make down the road from pollution control might be insufficient impetus to promptly clean up their act.

As far as ecological damage is concerned, some of the Bush people have convinced themselves that environmental threats are grossly exaggerated. Others consider nature resilient enough to withstand any of our abuse, or technology to be capable of repairing whatever injury we inflict on natural systems. There are a few (maybe more than a few) who think that accumulation of wealth, especially personal wealth, offsets environmental deterioration, rationalizing that if worse comes to worse, money can rescue anybody from any situation.

Bush has consistently demonstrated he believes a healthy environment is an automatic byproduct of a booming economy.

A healthy environment *can* be a byproduct of a robust economy, but it is *not* automatic. If you doubt this assertion, simply recall our own nation's compelling need 30 years ago to enact major anti-pollution statutes in the midst of prosperity.

There is no reason to back away from criticizing the Bush Administration's overzealous championing of private property rights. Those rights—sacrosanct as they might be—don't entitle individuals to purposely destroy natural resources that are part of the biological system upon which all life depends.

Faced with vocal opposition from environmental activists, whether in Congress or out in the hinterlands, the president tends to see red rather than green. That can complicate matters, but critics cannot afford to be intimidated when the fate of humanity hangs in the balance. Just keep name calling to a minimum and stress the shortcomings in Mr. Bush's thinking.

Abuse Of The Year

Want to know how much environmental damage a president can inflict in a single year? Take 2002. The Natural Resources Defense Council cataloged 150 instances of negative actions. In the interests of brevity (while still giving you an idea of the breadth of the Bush assault), a representative outrage has been selected for each month.

January—The White House oversees a weakening of the Clean Water Act regulations designed to protect wetlands from development.

February—President Bush unveils a global warming strategy that calls for increased study, defers emission reduction mandates, and would allow the main culprit—greenhouse gas pollutants—to continue to increase at a substantial rate.

March—The administration issues a permit to allow drilling for natural gas on Padre Island National Seashore in Texas, ignoring the destructive impacts on endangered sea turtles and other resident wildlife.

April—Under pressure from the White House, the U.S. Geological Survey reverses its conclusion that oil drilling in the Arctic National Wildlife Refuge could significantly harm wildlife.

May—Bush issues new regulations allowing coal companies to dump industrial waste from mountaintop mining into rivers, streams, lakes, and wetlands (a move subsequently blocked, at least temporarily, by environmentalists successfully petitioning a federal district court).

June—The president orders the Environmental Protection Agency to relax regulations that would force construction companies to reduce storm water runoff (the leading source of water pollution in the nation).

July—The administration opposes a provision in the Senate energy bill requiring utilities by 2020 to increase from two to 10 percent the amount of electricity they sell from clean, renewable energy sources. Such a shift to renewables would save consumers $13 billion and reduce smog.

August—For the first time ever, the White House permits energy development outside of already-leased areas within a national monument, specifically seismic exploration within the Ancients National Monument in Colorado.

September—It's revealed that the Bush Administration has reduced by 12 percent the number of personnel assigned to enforce the nation's air quality laws. The staff is at its lowest level on record.

October—The Superintendent of Yosemite National Park chooses to retire rather than be reassigned to the Great Smoky Mountains National Park in North Carolina, where the administration has ordered him to reverse longstanding government opposition to two environmentally harmful development projects.

November—Bush Administration officials announce that they are seeking ways to narrow the National Environmental Policy Act because the impact statements it requires are too time consuming. NEPA, the most important tool in staving off destructive environmental projects, is in jeopardy.

December—The president announces a wildfire suppression policy that environmentalists regard as a thinly veiled plan to accelerate logging of old-growth timber in remote sections of our national forests.

Is Anybody Listening?

Is anybody listening? It often doesn't seem so. The League of Conservation Voters, the lobbying arm of the national environmental movement awarded President George W. Bush a well-documented "F" for his mid-term environmental performance.

"Mr. Bush is well on his way to compiling the worst environmental record of any president in the history of our nation," declared Deb Callahan, LCV'S executive director, in releasing the assessment. Yet this highly incriminating indictment stirred barely a ripple.

Why has the public outcry been relatively muted, especially when compared to the furor over President Reagan's environmental policies, which never advanced to the degree Bush's initiatives already have?

Many Americans were traumatized by the World Trade Center terrorist attack, and President Bush's political strategy was to perpetuate that frame of mind. To a great extent, he was successful with both the electorate and the press. A high level of public anxiety diverted attention from Bush's efforts to institute controversial environmental reforms and other initiatives that lacked widespread societal support.

Even though he played the commander-in-chief role to the hilt, the president was aware that its magnetic public appeal had its limitations. Despite the favorable political climate, he did not attempt to pursue outright repeal of popular environmental regulations, as much as he might have desired to do so. Instead, his administration sought to roll back environmental laws in oblique, incremental, and (when possible) totally secretive ways in order to circumvent congressional debate and the media's glare. Favorite techniques included deliberately failing to fund or aggressively enforce popular environmental programs that the president disliked but did not dare cancel.

Where was the media during all of this subterfuge? While the press hammered the Reagan Administration for conservation faux pas, Bush's tenure was the backdrop for spotty, one might even say negligible, critical environmental coverage. Part of this had to do with preoccupation over terrorism and the Iraq War. Then there were editors who shied away from environmental stories because of their complexity and the lack of any imminent, dramatic resolution. Some newspapers were owned by corporate polluters, who whenever possible downplayed environmental coverage for fear of self-incrimination. Finally, especially in predominantly conservative Republican environs, there were some newspapers intimidated by the Hard Right's smearing of environmental activists as "extremists", "wackos," and "closet commies."

It is highly unlikely that Bush will backtrack to deflect criticism of his out-of-favor environmental policies. He genuinely believes his ideologically molded environmental approach is the correct one, even if the majority of Americans do not. Under his philosophy, nature is resilient enough to suffer plenty of abuse while being put in its proper place, namely in subservience to humanity. Pollution and paving over of valuable natural resources are the inevitable price of economic progress.

Conscious that many of his environmental views do not resonate with a majority of his fellow citizens, Bush bowdlerizes his radical policies with appellations that sound as "green" as any terminology a mainstream conservation organization would use.

Secret To Success

Presidential aspirants would be wise to ditch rigid ideological formulas and partisan considerations for practical solutions. That is the conclusion of a recent national survey of prospective voters conducted by the Opinion Research Corporation for the Newton, Mass.-based Civil Society Institute. The survey supported the private non-profit organization's view that the country would be better off with a bona fide give-and-take nationwide debate on major societal problems and an emphasis on pragmatic remedies.

Knee-jerk partisan rhetoric should be abandoned. Depending on the facts, the solution could turn out to involve free-market incentives so dear to many conservatives, or end up requiring the larger regulatory role that is often favored by liberals.

Nearly three-fourths of the survey participants—who represented a cross-section of Americans in age, gender, and political affiliation—criticized the nation's political leaders for failing to provide affordable health care for all citizens. Almost 60 percent blamed our leadership for falling short in applying practical solutions to the declining quality of public education and to persistent air pollution and other environmental problems.

> *Clearly, the candidate who rises above partisan mudslinging and strident ideological sound bites, engages in freewheeling forthright discussion of the issues, and elucidates common-sense solutions to major societal problems is going to attract voters.*

Clearly, the candidate who rises above partisan mudslinging and strident ideological sound bites, engages in freewheeling forthright discussion of the issues, and elucidates common-sense solutions to major societal problems is going to attract voters.

Conversely, candidates who are perceived to be manipulating the facts, constantly launching personal attacks, and insulating themselves from free and open debate will lose votes, especially among the uncommitted center of the American public that almost always determines a presidential election's outcome.

This has to be a concern for George W. Bush, who has been the most scripted president in memory. Bush's handlers have kept press conferences, media interviews, town hall meetings, and other forms of unrehearsed interaction with the public to below a bare minimum. It's not an ideal arrangement for a president who has been challenged on his forthrightness about Iraq and environmental reform. That is, of course, unless he has something to hide.

Fall From Grace

How the mighty have fallen! Thirty years ago, we were the world leaders in environmental protection, while such concerns were barely on the radar screen of our European allies.

Now, flash ahead to the present. In George W. Bush's most recent visit to England, our closest ally, his arrival was greeted by demonstrators waving placards reading "George Bush—wanted for environmental crimes against the planet."

And it's not just England that is critical of our environmental policies. Many nations in mainland Europe and elsewhere have been put off by our president's unwillingness to join international efforts to curb global warming.

Bush's environmental intransigence, along with his preemptive military invasion of Iraq, has resulted in a shocking decline in his image overseas. In a poll of Europeans conducted by the European Community in the autumn of 2003, Bush vied with North Korea's Kim Jong II as runner-up for the title of "head of state most threatening to world peace." (Israeli Prime Minister Ariel Sharon took top honors). [10]

While contemporary Europe has become a hotbed of environmental activism, the Bush Administration has been dragging its feet on the world stage. Almost overnight, we have become the laggards and the Europeans the progressives in addressing fundamental environmental challenges.

The role reversal is manifested in many ways. Europeans have eclipsed us in launching the transition to the mass application of clean, renewable energy. They have recently been more assertive in regulating toxic chemicals, and more willing to alert the public to the unknowns of genetically engineered food. Our friends across the Atlantic have been aggressive in setting deadlines and quotas for meeting pollution reduction goals. We have not.

To add insult to injury, even the developing world is beginning to upstage us. Hardly a paragon of environmental sensibility over the years, the Chinese government recently announced imposition of automobile fuel economy standards far tougher than our own.

Back in the mid-1970s, European political leaders were still preoccupied with the post-World War Two economic recovery. European environmental activists were lonely voices in a sea of indifference. They had no one to speak on their behalf in official circles, other than a handful of members of the European Parliament, which had little institutional authority in those days. There was no Environmental Protection Agency and limited opportunity to use the courts in the highly effective way American environmentalists have done. The only prominent European personages with any vision were a few elder statesmen, out of office, who foresaw the future environmental challenges but received more attention in America than in their homeland. In frustration, some French environmental activists actually imported the transcripts of our congressional hearings exploring the drawbacks of nuclear power and distributed the documents to every member of the National Assembly. Meanwhile, we were enacting one major environmental statute after another, gaining an enormous regulatory head start on our European brethren.

What is the explanation for the dramatic role reversal? Europeans' post-war apathy towards environmental concerns began catching up with them as pollution intensified with relatively unrestrained industrial activity. In response, they began emulating us in the past two decades, as environmental sensitivity took root among the public.

Meanwhile, the advent of the Reagan Administration marked a pullback from our environmental activism and aggressive regulatory policies of the 1970s. Especially in the case of Reagan and our current president, the highest priority has been to energize markets, even at the risk of gambling with public health. That hasn't played well of late with the Europeans, who lack the Bush White House's confidence that a largely deregulated marketplace can meet society's utilitarian demands. Our allies also don't take kindly to Bush's unilateral rejection of a number of international environmental pacts. To many, it looks as though the United States considers itself above the laws that apply to everybody else.

Suddenly, the European "students" find themselves surpassing their American "teachers" in environmental leadership. Our fall from "environmental grace" is certainly a cause for embarrassment but, given the adverse global ramifications of our regression, it is even more a cause for concern.

The Worst Ever?

George W. Bush has emerged as the worst environmental president in the nation's history, surpassing the previous titleholder, Ronald Reagan. That is the conclusion of the national environmental community, and here is why.

Bush is far more of a "green" menace than Reagan because of the ability to operate in a manner that does not attract attention. Reagan's approach was blatant, and therefore relatively easy for environmentalists to expose and head off at the pass.

Bush has also been helped by the fact that his party controls both houses of Congress, and The GOP leadership shares his conservative ideological bent.

President Reagan was quite detached from environmental issues, so when his initiatives were rejected, he rarely chose to put up a fight. Not so George Bush. He has proven to be much more of an activist in pushing a conservative environmental agenda in which the private sector and corporate voluntary compliance take precedence over the public sector and governmental regulation.

Bush has opened up only three new national parks, fewer than any president in the past 100 years, and established the least number of acres of wilderness of any White House occupant since the 1964 Wilderness Act was signed into law. The president requested only $425 million for international family planning programs in 2005, more than a 35 percent reduction from a decade earlier, even though global population has increased by half a billion people over the same period of time.

While he has occasionally beaten a retreat when one of his proposals has caused too much of a furor, his concessions have tended to be cosmetic rather than substantive. Bush knows many Americans don't share his convictions, which is why he has resorted to deceptive rhetoric, arcane procedural methods, and budgetary sleight of hand. It is this pernicious Fifth Column approach that has led environmentalists to adopt the mindset, come November 2004, "anybody but Bush."

Chapter Nine

Where To From Here?

Future Challenges

In the broadest sense, the 21st Century's greatest challenge will be whether the human race can rescue the earth's beleaguered natural systems from ever-increasing stress. Success will be measured by the extent to which mankind can make the transition to an environmentally sustainable lifestyle.

Failure to meet this challenge will exacerbate the major environmental threats confronting us at the start of a new millennium: incremental pollution of the air, water, and soil; potential shortages of fresh water in arid regions; global warming; loss of biodiversity and fertile cropland; and runaway population growth that intensifies the aforementioned problems.

> *The new millennium will also test our ability to manage with minimal disruption the inevitable passage from a fossil fuel-dependent global community to one that relies primarily on clean, renewable sources of energy.*

The new millennium will also test our ability to manage with minimal disruption the inevitable passage from a fossil fuel-dependent global community to one that relies primarily on clean, renewable sources of energy. Another significant challenge will be getting a handle on genetic engineering before it gets a handle on us.

Without population stabilization, the earth's already depleted natural resource base will be subject to further pressures that would increase international tensions and invite domestic strife.

Wise land-use planning must become universal, so that areas with the greatest agricultural productivity and biodiversity can be preserved for future generations. There is some urgency here. According to the latest United Nations Food and Agriculture Organization's report, world hunger is increasing after declining steadily during the first half of the 1990s. There are already an estimated 842 million undernourished people on the planet, and though that may be more a problem of distribution than supply, further loss of farmland could result in an even more precarious situation.

To achieve sustainability, no more daunting task awaits humanity than the need to alter a value system that crowns material acquisition as the dominant measure of success and principal formula for happiness. The recommended alternative is not to exchange comfort for deprivation. Rather it is to attach greater prestige to constructive personal achievement and compilation of knowledge than to the accumulation of a vast quantity of material possessions. It means placing greater emphasis on quality over quantity of goods and services. Recycling and reuse of materials would be integrated into the daily routine to a far greater extent than occurs now.

Another essential step would be to codify the principle that public and environmental health takes precedence over profit. Without adhering to that doctrine, altruism would be at the perpetual mercy of avarice, and the future would take on a decidedly ominous cast.

Ask yourself whether George W. Bush is up to these challenges.

An Ecumenical Test

Has the war on terrorism caused President George W. Bush to rethink his reflexive, highly ideological stance toward many global environmental problems and adopt a more ecumenical, pragmatic approach?

In the aftermath of the terrorist attack, Bush did reconsider his withdrawal from deliberations involving an international ban on biological warfare. That was only one of a number of treaties the president snubbed in a fit of unilateral pique prior to the September 11 destruction of the World Trade Center.

Many of the countries smarting at the Bush Administration's go-it-alone policy were the very ones the president was soliciting for his coalition against worldwide terrorism. Environmentalists had hoped that, in order to mollify the aforementioned nations and achieve the support he sought, the president would reverse field and rejoin the international community to negotiate a host of environmental treaties that he had previously spurned.

So far, they have been disappointed. Our isolation on the international stage continues.

Wild Card

How would environmental issues fare under Bush if we were to experience another horrendous terrorist attack? Chances are they would disappear from the president's radar screen, just as they did in the immediate aftermath of 9/11. Bush would delegate environmental policy totally to his subordinates, most of whom are conservative ideologues committed to the pro-industry tilt of Bush's unpopular pre-9/11 environmental policy. The question that then would arise is whether an emotional American public, in its eagerness for national unity, would give the Bush Administration pretty much a free pass on the domestic front, as it briefly did after the World Trade Towers went down. My guess is a blank check would be granted, but of a far shorter duration than the 9/11 honeymoon. The shock would wear off more rapidly, because the pre-9/11 element of surprise no longer exists.

The terrorist attacks have sobered and matured Bush's handling of foreign policy to a great extent. But there is little evidence to date that the overseas threat has caused him to reassess his provincial views on environmental protection.

The World Under Bush

Welcome to a world where large cities are routinely choked by lung-irritating photochemical smog and clogged with ever-increasing traffic congestion. If you are living in an American metropolis and don't telecommute, your automobile trip to and from work averages nearly an hour each way. In some major urban centers, the downtown areas are so gridlocked that motorists can enter by car only on alternate days.

It's a world where wilderness landscapes are dotted with oil derricks. Indeed, nature is in full retreat from virtually unrestrained development. Overseas, Australia's Great Barrier Reef is the planet's last surviving major coral reef system and is steadily shrinking, as ocean temperatures continue to rise in conjunction with global warming. Due to overfishing, many popular species no longer can be found in the supermarket, and those on display are priced beyond the reach of all but the wealthiest Americans.

To keep from being inundated by the rising level of the Atlantic Ocean, portions of the United States' East Coast have replicated Holland's elaborate dike system. In the northern tier of the United States, thunderstorms are more likely than snowstorms to occur in the dead of winter. New York City has air-conditioned its armories and other public buildings to serve as long-term emergency shelters for senior citizens otherwise unable to escape prolonged heat waves. Other major American cities are following suit. All professional sporting events during the summer must take place in indoor, air-conditioned arenas, due to inhospitable temperatures and air quality. City dwellers routinely wear air filter masks outdoors during the most pollution-prone months of the year.

Birth rates and maternal mortality are soaring where reproductive health services and sex education have been severely restricted because of the U.S. government withholding financial aid in order to discourage the "immorality" of abortion counseling and other family planning programs.

Synthetic products are replacing the waning supply of many different raw materials that we failed to recycle.

Opportunities have practically disappeared for the American public to comment on government environmental rulemaking or mount a legal challenge against a pending decision deemed objectionable. Meanwhile, industry lobbyists' influence on federal environmental policy is at an all-time high.

Welcome to George W. Bush's world, for this is where we are heading if the president's largely do-nothing response to global warming, myopic approach to family planning, penchant for secrecy, and other regressive environmental policies become entrenched in our society.

A Question Of Values

The fate of the vast environmentally degraded swamp in southern Iraq remains a potentially major sticking point in the aftermath of Saddam Hussein's ouster.

Prior to being devastated by Hussein, this terrain was a biologically rich wetland that was home to as many as 500,000 Iraqi Shiites. These people—known as "marsh Arabs"—had lived for centuries in relative pastoral tranquility on floating islands constructed from reeds, and earned their keep largely from fishing and farming. They were surrounded by an ecosystem larger than the Everglades and teeming with wildlife, including a number of endangered species. It also served as a valuable water storage area and filter for purifying the flow of Iraq's rivers into the Persian Gulf.

In the aftermath of the 1991 Gulf War, Saddam crushed the Shiites' revolt, drove them from their homes, and drained most of the swamp, leaving only seven percent of the original ecosystem intact.

With the removal of Saddam from power, a sizable cadre of international planners is hoping to flood a large portion of the marshes and make the region habitable once again.

But not so fast! Another scenario recently has emerged and is being promoted for the Mesopotamia marshlands. Some geologists believe there are vast oil deposits underneath this ecosystem.

What will be the region's fate? When the energy industry contemplates the area, it doesn't envision lush foliage, the return of clean water, a profusion of migratory birds, the salvation of endangered species, and the restoration of a pastoral civilization. It fantasizes about derricks, pumps, and pipelines crisscrossing the landscape, and most of all, the handsome profits to be made.

Iraq's vast untapped oil fields outside the Mesopotamia swamp are evidently not enough of a treasure trove. In the industry's universe, greed apparently conquers all.

So what will it be? Oil or water? As the principal occupier of Iraq in the immediate aftermath of the war, the United States government has a major say.

In which direction will the Bush Administration lean? I would like to think that the White House would opt for restoration of an environmentally sustainable civilization as the most humane thing to do, as well as the most ecologically prudent course for a "liberated" Iraq to take. But the greatest utility of a landscape is in the eye of the beholder, and the Bush Administration's history in defining what constitutes the highest value of land cannot be reassuring to the Marsh Arabs yearning to return home.

To demonstrate what I mean, note the following descriptions of the Arctic National Wildlife Refuge's coastal plain delivered at a recent congressional hearing. The first version was articulated by Bush's Interior Department Secretary, Gail Norton, who was trying to convince lawmakers it was perfectly acceptable to drill for oil in the midst of this unique Alaskan wilderness. The second was by National Wildlife Federation official Jamie Rappaport Clark, former head of the U.S. Fish and Wildlife Service.

Norton told the legislators that "we call the area in dispute the coastal plain because it is just that—a plain. There are no trees, there are no deepwater lakes. There are no mountains like those in an environmental group's video. Outside the coastal plain, there are mountains in ANWR, but they are designated as wilderness areas, and no one is remotely considering them for energy production . . . Now let's take a look at what the coastal plain of Alaska actually looks like most of the year.

[She then displayed a grim-looking video produced by Arctic Power, a pro-drilling group.]

"This is what I saw when I was there in the last day of March 2001, with a 75-degree-below-zero wind chill. This image of flat, white nothingness is what you would see the majority of the year. In fact, there are 56 days of total darkness during the year, and almost nine months of harsh winter."

Now for the picture that Ms. Clark painted: "No other conservation area in North America safeguards a complete range of Arctic and sub-Arctic ecosystems. No other, in the entire five-nation circumpolar north, has as abundant or diverse wildlife. Even when locked in the frigid grip of winter, the coastal plain is never lifeless. Musk oxen, cloaked in shaggy wool, restrict their movements to conserve vital energy reserves. Hidden from view, maternal polar bears give birth and nurse their young in the thermal protection of snow dens. Arctic foxes and ptarmigan—predator and prey—camouflage in winter white coats. Fish like Arctic grayling and Dolly Varden survive in rare pockets of open water beneath the ice-covered rivers and lakes. In late spring, the coastal plain transforms, as do few places on earth . . . Caribou have already begun their annual trek northward across the Brooks Range to this place that has served as their central calving and nursery ground for thousands of years. From continents away, flocks of migratory birds are on wing to the coastal plain, which, by summer, will be filled with a symphony of bird songs. Arctic foxes, red foxes, grizzly bears, and wolverines will thrive and fatten amid this abundant life."

It's hard to believe these two individuals were talking about the same piece of land. But they were. That bodes ill for the Marsh Arabs' future, considering that the Bush Administration is taking the lead in overseeing postwar Iraq and could well impose its world view on the Mesopotamia wetlands through its oil-tinted glasses.

Power Failure

Recent events highlight some of the flaws in the Bush Administration's vision for this country.

For example, the administration hopes to parlay the Northeastern United States blackout into a justification for oil drilling in the Arctic National Wildlife Refuge's pristine coastal plain. But the cause of the power failure was a distribution problem, not a supply one, and the greatest concern emerging from the crisis involves the national electricity grid's vulnerability to terrorist sabotage.

Then there is George W's acceptance as gospel that land-use planning is a local matter, and that any intervention by the federal government is apt to smack of heavy-handed intrusiveness associated with a totalitarian state.

Yet a recent University of Maryland study warned that local zoning decisions that allow undeveloped green space to be converted into cities and farms could be contributing twice as much to global warming as originally thought. It just may be that the national land-use planning that Bush abhors will be necessary to save us from ourselves.

The Vision Thing

Let's talk about visions for the future of the nation, the "vision thing," so to speak.

> The long-term outlook is not especially promising if
> mankind continues to utilize the earth's natural resources
> at a rate exceeding the planet's regenerative capacity.

The long-term outlook is not especially promising if mankind continues to utilize the earth's natural resources at a rate exceeding the planet's regenerative capacity. President George W. Bush doesn't seem to have a feel for either the problem or the answers. His vision of the world appears tied to perpetuating conspicuous consumption as society's mantra, allowing oil rigs to mar the wilderness landscape, and requiring environmental protection to take a back seat to economic growth and militarization.

Note that the estimated military expenditures for the entire world in 2003 amounted to more than $900 billion. If only 12 percent of that sum could have been siphoned off, it would have been enough to halve global poverty and hunger levels by the year 2015, as well as meet most of the human race's health and education requirements.

Without this shift in priorities, approximately 24,000 people will continue to die every day from hunger, 12 million individuals will succumb each year from lack of clean water, and more than half of humanity will continue to struggle to survive on less than two dollars a day.

Unfortunately, Bush's vision of the future offers scant opportunity for these poverty-stricken individuals to improve their lot and, even in a larger sense, fails to supply a framework for getting the human race back on a sustainable track.

Not Getting It

The way President Bush operates, it's hard to believe he is aware that less than a quarter of the earth's land mass remains completely undeveloped, with the bulk of that territory consisting of inhospitable desert, mountains, and permanently frozen terrain. If he is cognizant that most of the world's productive land has been or is being stripped of its biological diversity by humanity, he has a pretty odd way of showing it. Under his sway, we have been dragging our feet on joining the international treaty to preserve that very biological diversity.

Even when we do sign on to an environmental pact, we can be finicky. Note that the head of our delegation at an international conference warned we might repudiate the landmark Montreal Protocol treaty if our farmers were not granted an exemption to use the pesticide methyl bromide. Hopefully, it is an empty threat. This treaty was created to halt destruction of the earth's protective ozone layer by such manmade chemical compounds as methyl bromide rising into the stratosphere—and the pact has been effective in reducing harmful emissions.

American farmers, whose crop yields would decline without methyl bromide, ought to receive government assistance to tide them over until they have access to a suitable substitute. But under no circumstances should they be allowed to flout international law, much less jeopardize the health of the atmosphere for present and future generations. Any economic hardship to the farmers, when measured on a utilitarian scale, is the lesser of two evils.

Unraveling The Base?

Bush junior's environmental policies are alienating some of his core constituencies and quite possibly driving them into the arms of his political adversaries.

His plans to promote massive expansion of oil and gas drilling in the Rocky Mountain region has some of his traditional allies—local ranchers, farmers, hunters, and sport fishermen—up in arms. Normally suspicious of environmental activist organizations, these staunch Republican outdoors types have suddenly found themselves aligned with the "green" groups. Their shared fear is that the energy industry will run roughshod over the land that provides sustenance for livestock, habitat for wildlife, and glorious scenery as a tonic for the human eye and spirit.

Whether this will translate into these avowed Republicans deserting George Bush on Election Day remains to be seen. For some, it will undoubtedly depend on whether they consider John Kerry a viable alternative.

Consumed With Consumption

Given our culture's allegiance to conspicuous consumption, any call to lighten our impact on the earth's beleaguered natural systems would seem destined to fall on a lot of deaf ears.

We certainly have an ardent champion of unbridled material acquisition currently residing in the White House. One of President Bush's first responses to the September 11th terrorist attacks was to urge Americans to defy the enemy by descending on shopping malls and spending like crazy. His exhortation can be traced to a conviction pervasive throughout our history that the fate of the nation's economy depends upon the *level* of consumption. In point of fact, the future health of our country hinges on the *kind* of consumption in which we engage, namely the type that doesn't undermine the regenerative capacity of natural systems.

Unfortunately, recognition that consumption ought to be a means to an end rather than an end in itself has yet to register for many in our society. A "greed is good" mentality is pervasive and fosters environmental degradation.

We now have more vehicles than people licensed to drive them. By contrast, 40 percent of the world's people have yet to reach a consumption level that consistently provides the basic necessities!

Somehow, we must come to realize we need to set an example for the rest of the world by scaling back our environmentally unsustainable consumption. At the same time, we have to convince developing countries not to emulate our current unrestrained materialism once they can provide all their people with the necessary creature comforts.

These changes in direction are a mix of equity and environmental imperative. Besides reducing the enormous disparities in standards of living, they would go a long way toward preventing further deterioration of the earth's overtaxed ecosystem and reversing the damage that has been done.

Chances would appear bleak for us to abandon our spendthrift ways, at least in the near term. It's hard to imagine a more hostile political climate for such reforms than the one that exists now under President Bush. Could you imagine him supporting the imposition of an entirely new eco-tax on goods and services in order to discourage consumption that places an undue strain on the biosphere? Even if such a levy failed to deter a determined big spender, at least revenues could be raised to repair environmental despoliation or, better yet, finance protection that would prevent any damage in the first place.

An acquisitive temperament is too much a part of human nature to ever be purged, but that is no fatal flaw. A sense of luxury can be created in an environmentally sound manner. The trick is to wean oneself from the idea that success is defined by excess, and that moderation is a portal to deprivation.

Signs Of Hope

All is not doom and gloom on the American scene, despite our environmentally unsympathetic national leadership. Let me cite several examples of positive developments.

There is hope that future economic benefits from undeveloped public lands will be derived in a more environmentally friendly way than the large-scale denuding of natural resources that goes on now. The Interior Department recently released a report that found approximately 66 million Americans spend more than $38 billion annually observing, feeding, or photographing wildlife in our public parks and other wilderness sanctuaries. For each dollar that wildlife watchers spend on lodging, food, and transportation, as well as on such items as cameras, film, and binoculars, the report estimates a ripple effect producing a total yearly industrial output worth $95.8 billion. That includes the creation of more than one million jobs, with total wages and salaries of $27.8 billion. [11] A cottage industry to replace drilling and mining of federal lands is fast emerging.

Then there is the grassroots syndrome. If progressive leadership doesn't emanate from the top down, there is always the chance that it will flow from the bottom up. Note the recent action taken by twelve states, including California and New York. They filed petitions in federal court to compel a stubbornly resistant Bush Administration to regulate emissions of the greenhouse gas carbon dioxide. The petitions asked the judge to review an Environmental Protection Agency decision that it did not have authority to regulate such emissions under the Clean Air Act (a transparent excuse to wiggle out of putting the compliance squeeze on George W. Bush's corporate cronies). It is comforting to know that if Washington is derelict in meeting its responsibilities, there are at least some states prepared to pick up the slack. [12]

Earth Day 2004

The arrival of Earth Day 2004 generated the usual flurry of "feel good" stories. While they ought not be ignored, Earth Day at this juncture in our history should be more of a reminder than a celebration.

That is clear from a set of statistics generated by the Environmental Protection Agency and other government sources. According to the documentation, the nation's water pollution levels are increasing for the first time since passage of the 1970 Clean Water Act. Our estuaries have deteriorated, with more than half "impaired"—up from 37 percent in 1994. There are 218 million Americans now living within 10 miles of a body of water unfit for drinking, fishing, swimming, or boating.

The U.S. Centers for Disease Control reports that one out of 12 women of childbearing age in this country has unsafe mercury levels in her body, with the result that more than 320,000 newborns are subject to severe health risks annually.

Automobile fuel efficiency has sunk to its lowest level in two decades and contributed to more air pollution from vehicular emissions, as well as increased the political pressure to open up heretofore protected public lands to oil drilling.

Fish are too contaminated in 28 states, 82 percent of our estuaries, and along the majority of the coastline in the lower 48 states to be eaten regularly.

Lake Erie's dead zone is increasing for the first time in 30 years. A quarter of Americans live within four miles of an active Superfund hazardous waste site. Most large oceanic fish populations have plummeted to 10 percent of their 1950 levels.

Add to the mix the growing concerns over global warming, a human population explosion, increased poverty, the accelerated rate of species extinction, and a precipitous loss of biologically rich tropical forests and other prime wildlife habitat to development.

The conclusion should be obvious, as should the message of Earth Day 2004. Now is not the time for complacency (a recent poll shows public concern about the environment has declined). The challenge to restore our planet to full biological health is greater than ever.

That is the environmental backdrop for a national referendum on whether George W. Bush deserves a second presidential term.

Credit Where Credit Is Due

As this book goes to press, the Bush Administration has issued a strong anti-pollution rule to regulate diesel fuel emissions of off-road vehicles. For the first time in the president's tenure, he has won universal acclaim from the environmental community for an initiative.

The issue now becomes whether this action is an election-year maneuver to mute criticism of his past record or represents a genuine change of heart. Given George W. Bush's history, the extent of his conversion remains open to question until his commendable initiative no longer stands in conspicuous isolation.